STUDY GUIDE
for Moore's

The Basic Practice of Statistics

Third Edition

Michael A. Fligner
William I. Notz
Both of Ohio State University

W. H. Freeman and Company
New York

Printed in the United States of America

ISBN: 0-7167-5886-5

First printing 2003

W. H. Freeman and Company
41 Madison Avenue
New York, NY 10010
Houndmills, Basingstoke
RG21 6XS, England

CONTENTS

CONTENTS

CHAPTER 1

PICTURING DISTRIBUTIONS WITH GRAPHS

OVERVIEW

Understanding data is one of the basic goals in statistics. To begin, identify the **individuals** or objects described, then the **variables** or characteristics being measured. Once the variables are identified, you need to determine whether they are **categorical** (the variable puts individuals into one of several groups) or **quantitative** (the variable takes meaningful numerical values for which arithmetic operations make sense). The guided solution for Exercise 1.1 given below provides more details on deciding whether a variable is categorical or quantitative.

After looking over the data and digesting the story behind it, the next step is to describe the data with graphs. The first methods are simple graphs that give an overall sense of the pattern of the data. Which graphs are appropriate depends on whether the data are numerical or not. Categorical data (nonnumerical data) use **bar charts** or **pie charts**. Quantitative data (numerical data) use **histograms** or **stemplots**. Quantitative data collected over time use a **timeplot** in addition to a histogram or stemplot.

When examining graphs be on the alert for

- **outliers** (unusual values) that do not follow the pattern of the rest of the data
- some sense of a **center** or typical value of the data
- some sense of how **spread** out or variable the data are
- some sense of the **shape** of the **overall pattern**

In timeplots, be on the lookout for **trends** over time. These features are important whether we draw the graphs ourselves or depend on a computer to draw them for us.

GUIDED SOLUTIONS

Exercise 1.1

KEY CONCEPTS - individuals and type of variables

a) When identifying the individual or objects described, you need to include sufficient detail so that it is clear which individuals are contained in the data set.

b) Recall that the variables are the characteristics of the individuals. Once the variables are identified, you need to determine if they are categorical (the variable puts individuals into one of several groups) or quantitative (the variable takes meaningful numerical values for which arithmetic operations make sense). Think carefully about how you want to treat the "Number of Cylinders." Now list the variables recorded and classify each as categorical or quantitative.

Name of variable	Type of variable

Exercise 1.5

KEY CONCEPTS - drawing a histogram

Hints to remember in drawing a histogram:

1. Divide the range of values of the data into classes or intervals of equal length.
2. Count the number of data values that fall into each interval.
3. Draw the histogram.

   a) Mark the intervals on the horizontal axis and label the axis. Include the units.

   b) Mark the scale for the counts or percents on the vertical axis. Label the axis.

   c) Draw bars, centered over each interval, up to the height equal to the count or percent. There should be no space between the bars (unless the count for a class is zero, which creates a space between bars).

For the exercise:
The vertical axis in a histogram can be either the count or percent of the data in each interval. Changing the units from counts to percents will not affect the shape of the histogram. Complete the table of the number of cars in each miles per gallon class. The first class is done for you. With this small number of data values it is both easy and instructive to draw the histogram by hand.

Miles per gallon distribution

Class	Count
10 to 14 mpg	1
15 to 19 mpg	
20 to 24 mpg	
25 to 29 mpg	
30 to 34 mpg	
35 to 39 mpg	
40 to 44 mpg	
45 to 49 mpg	
50 to 54 mpg	
55 to 59 mpg	
Total	

The values of the variable mpg have gaps. To avoid the gaps in the histogram, the bases of the bars need to be extended to meet halfway between two adjacent values. For example, the bar representing 20 - 24 mpg would need to meet the bar from 25 - 29 mpg at 24.5 Similarly, the bar representing 25 - 29 mpg would need to meet the bar from 30 - 34 mpg at 29.5. Thus the bar representing 25 - 29 mpg must go from 24.5 to 29.5, with a center at 27. Each bar has a width of 5 mpg as required in this exercise. In the histogram given, the *centers* of the bars are provided on the horizontal axis, and the first bar has been completed for you.

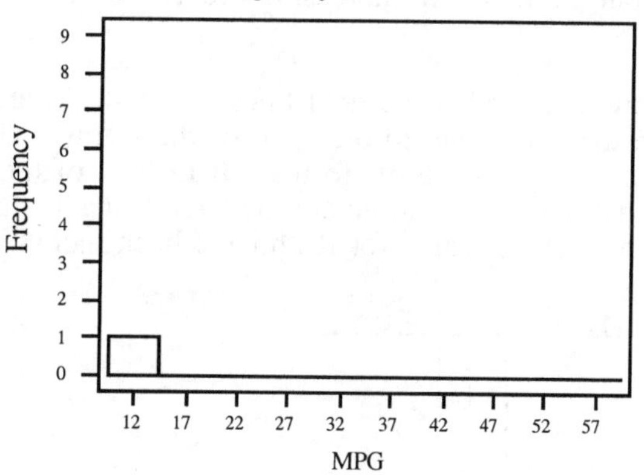

Exercise 1.11

KEY CONCEPTS - drawing bar charts

What is the total of the percents in the table? What percent do you think have other colors? Would a pie chart be appropriate? Complete the bar chart below. The first bar has been completed for you.

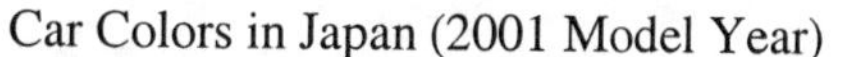

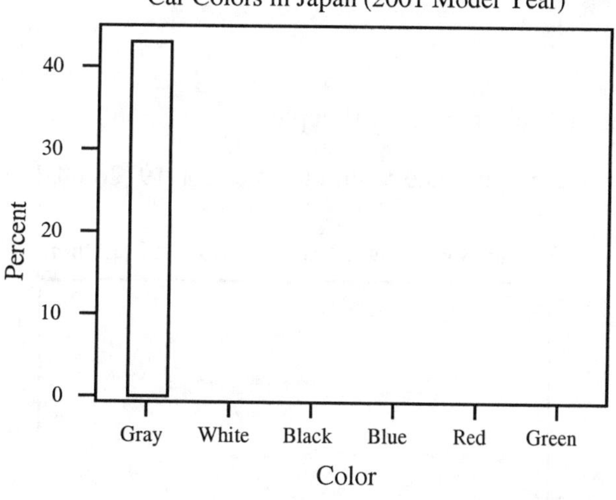

What do you see as the most important differences between choice of color in Japan and North America (See Exercise 1.3 in your text).

Exercise 1.26

KEY CONCEPTS - drawing stemplots, back-to-back stemplots, comparing two stemplots

Hints to remember for drawing a stemplot:

1. Put the observations in numerical order.

2. Decide how the stems will be shown. Commonly, a stem is all digits except the rightmost. The leaf is then the rightmost digit.

3. Write the stems in increasing order vertically. Write each stem only once, unless you are splitting the stems. Draw a vertical line next to the stems.

4. Write each leaf next to its stem.

5. Rewrite the stems and put the leaves in increasing order.

For the exercise:
We have given the stems below, enclosed in vertical lines. The stems are in units of 10 home runs. Make the stemplot for Ruth in the usual way to the right of the stems. The stemplot for McGwire is made in the same way, with the leaves going off to the left instead of the right. This type of plot is excellent for comparing two small data sets. Complete the back-to-back stemplot below. The smallest two values for McGwire and the smallest value for Ruth have been included to help you get started.

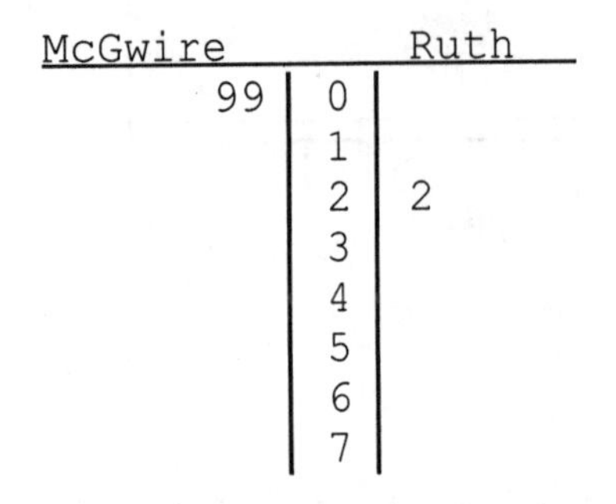

How do the two players compare? Concentrate on center, spread, outliers and overall shape of the two stemplots. Which points do you think correspond to 1993 and 1994 for McGwire?

Exercise 1.27

KEY CONCEPTS - drawing and interpreting a timeplot

a) Complete the timeplot on the graph. The winning times for 1972 and 1973 are included in the plot to get you started.

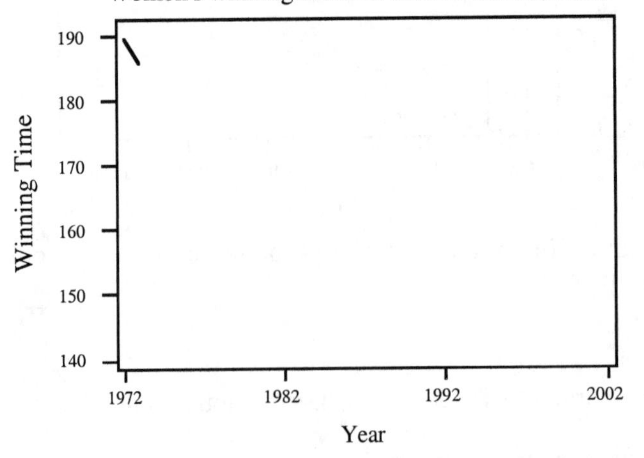

b) What is the general pattern in the timeplot? Have times stopped improving in recent years?

COMPLETE SOLUTIONS

Exercise 1.1

a) The individuals in this exercise are the different make and model cars.

b) The variables are vehicle type (categorical), transmission type (categorical), number of cylinders (probably can be considered as categorical), city MPG (quantitative) and highway MPG (quantitative). Although the variable cylinders is recorded as a number, this variable divides the cars into only a few categories. For a group of cars, we would not necessarily compute the average number of cylinders, but instead the number or percentage of 4-, 6- or 8-cylinder cars.

Exercise 1.5

The distribution of highway miles per gallon for two-seater cars and the corresponding histogram are given below. Most cars get between 20 and 30 mpg on the highway with the Honda Insight at 56 mpg being a clear outlier.

Miles per gallon distribution

Class	Count
10 to 14 mpg	1
15 to 19 mpg	2
20 to 24 mpg	6
25 to 29 mpg	9
30 to 34 mpg	3
35 to 39 mpg	0
40 to 44 mpg	0
45 to 49 mpg	0
50 to 54 mpg	0
55 to 59 mpg	1
Total	22

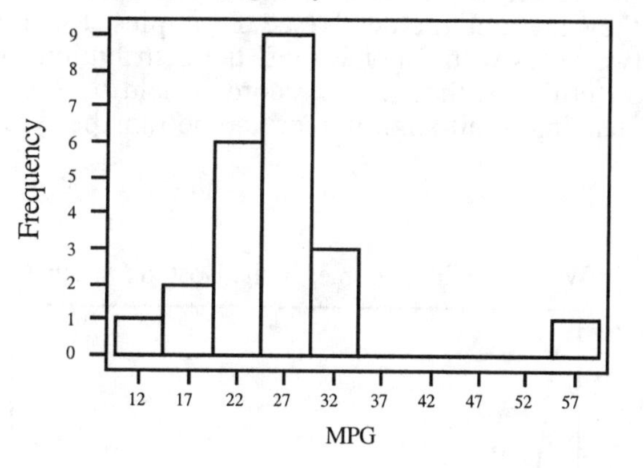

Exercise 1.11

The percents in the table add to 99%. It could be that 1% of Japanese vehicles are made in other colors, or the fact that the percents add to only 99% may be due to roundoff error. A pie chart could be drawn without including an "other" category in this case. The most important difference in vehicle colors in Japan and North America is there are substantially fewer car color choices in Japan.

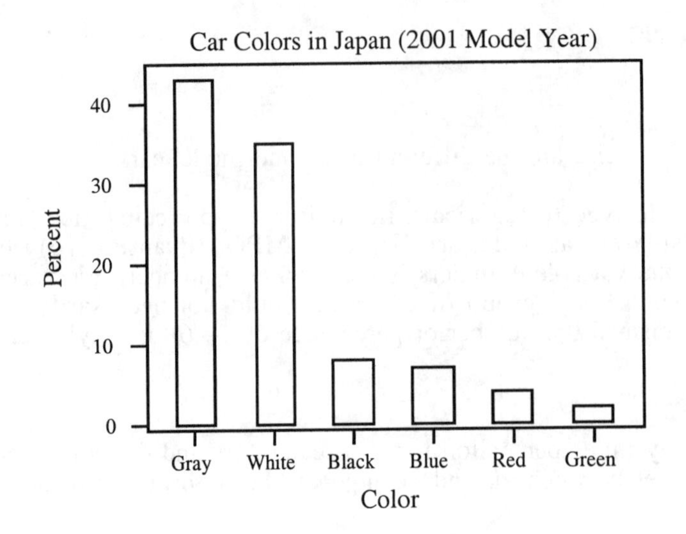

Exercise 1.26

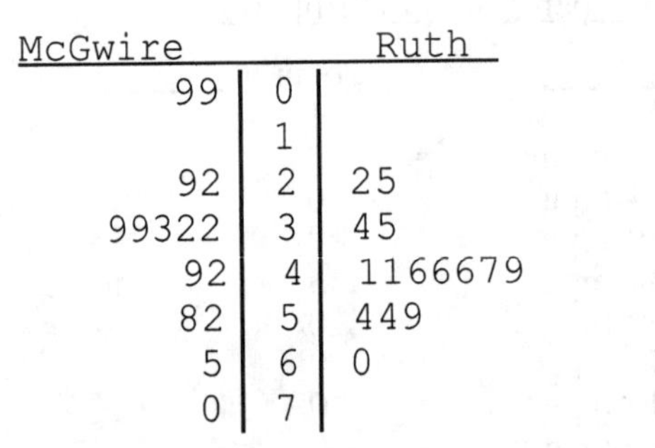

McGwire		Ruth
99	0	
	1	
92	2	25
99322	3	45
92	4	1166679
82	5	449
5	6	0
0	7	

Ruth's distribution is centered at 46 home runs per year and looks fairly symmetric. His famous record of 60 home runs in 1927 does not appear to be an outlier, given his record.

The year in which McGwire was injured and the year of the baseball strike each have 9 home runs. While they show up as outliers, there are good explanations for both. We might want to exclude these years as not full seasons. The center of McGwire's home run distribution is 39 including the two outliers, and is still 39 if they are not included (see Examples 1.4 and 2.2 in the text for the calculation). Excluding the two years with 9 home runs, the distribution looks slightly skewed to the right with a similar spread to Ruth's. Although the record is held by McGwire, a comparison of the two stemplots shows Ruth as tending to hit slightly more home runs per year.

Exercise 1.27

a)

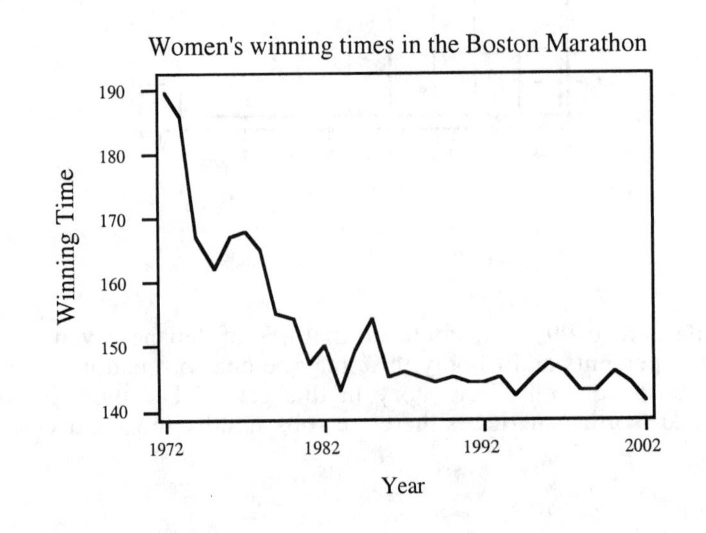

b) There is a fairly steady downward trend in the winning times until around 1982. After this point the improvement seems to have stopped with the winning times varying around 145 minutes. The variability in the winning times from year to year has also gone down.

CHAPTER 2

PICTURING DISTRIBUTIONS WITH NUMBERS

OVERVIEW

Once you examine graphs to get an overall sense of the data, it is helpful to look at numerical summaries of features of the data that clarify the notions of center and spread.

Measures of center:
- **mean** (often written as $\bar{x}$)
- **median** (often written as M)

Finding the mean $\bar{x}$

The mean is the common arithmetic average. If there are n observations, $x_1, x_2, \ldots, x_n$, then the mean is

$$\bar{x} = \frac{x_1 + x_2 + \ldots + x_n}{n} = \frac{1}{n}\sum x_i$$

Recall that $\sum$ means "add up all these numbers."

Finding the median M

1. List all the observations from smallest to largest.
2. If the number of observations is odd, then the median is the middle observation. Count from the bottom of the list of ordered values up to the $(n + 1)/2$ largest observation. This observation is the median.
3. If the number of observations is even, then the median is the average of the two center observations.

Measures of spread:
- **quartiles** (often written as Q_1 and Q_3)
- **standard deviation** (s)
- **variance** (s^2)

Finding the quartiles Q_1 and Q_3

1. Locate the median.
2. The first quartile, Q_1, is the median of the lower half of the list of ordered observations.

3. The third quartile, Q_3, is the median of the upper half of the list of ordered values.

Finding the variance s^2 and the standard deviation s

1. Take the average of the squared deviations of each observation from the mean. In symbols, if we have n observations, $x_1, x_2, \ldots, x_n$ with mean $\bar{x}$

$$s^2 = \frac{(x_1 - \bar{x})^2 + (x_2 - \bar{x})^2 + \ldots + (x_n - \bar{x})^2}{n-1} = \frac{1}{n-1}\sum(x_i - \bar{x})^2$$

(Remember, Σ means "add up.")

If you are doing this calculation by hand, it is best to take it one step at a time. First calculate the deviations, then square them, next sum them up, and finally divide the result by $n - 1$.

2. The standard deviation is the square root of the variance, i.e., $s = \sqrt{s^2}$. Some things to remember about the standard deviation:

a) s measures the spread around the mean.

b) s should be used only with the mean, not with the median.

c) If $s = 0$, then all the observations must be equal.

d) The larger s is, the more spread out the data are.

e) s can be strongly influenced by outliers. It is best to use s and the mean only if the distribution is symmetric or nearly symmetric.

For measures of spread, the quartiles are appropriate when the median is used as a measure of center. In fact, the **five-number summary**, reporting the largest and smallest values of the data, the quartiles, and the median, provides a compact description of the data. The five-number summary can be represented graphically by a **boxplot**. If you use the mean as a measure of center, then the standard deviation and variance are the appropriate measures of spread. Watch out, because means and variances can be strongly affected by outliers and are harder to interpret for skewed data. The mean and standard deviation are not resistant measures. The median and quartiles are more appropriate when outliers are present or when the data are skewed. The median and quartiles are **resistant measures**.

GUIDED SOLUTIONS

Exercise 2.7

KEY CONCEPTS - stemplots, split stems, mean and five-number summary

Complete the stemplot of the ages of the presidents at their inauguration provided on the next page. If you used only three stems, one for the ages in the 40s, one for those in the 50s, and the last for those in the 60s, most of the features of the distribution would be obscured. In this case splitting the stems is helpful: split each of these stems into two stems. The first stem starting with 4 would have leaves 0 through 4, the second stem starting with 4 would have leaves 5 through 9, and so forth. Fill in the leaves for the following stemplot. What is the shape of the distribution? Are there outliers? How does this affect the relationship between the mean and the median?

```
4 |
4 |
5 |
5 |
6 |
6 |
```

b) The mean is most easily computed on a calculator. The sum of all the ages is 2357. You can use this to evaluate

$\bar{x}$ =

Complete the five-number summary for the distribution of age at inauguration. It's simplest if you first list the ages in increasing order below. There are many "ties" in the data, so be careful when following the rules for finding quartiles.

<u>Five-number summary</u>
Minimum
First quartile
Median
Third quartile
Maximum

c) The middle half of the ages falls between which two numbers in the five-number summary? What is the range of these numbers?

The information in the five-number summary can be used to describe the ages of the youngest 25%. Was Bill Clinton in this group?

Exercise 2.9

KEY CONCEPTS - stemplots, mean and standard deviation, shape of distribution

Use statistical software or a calculator to find the mean and standard deviation of the two data sets and fill in your answers below.

Data A	$\bar{x}$ =	s =
Data B	$\bar{x}$ =	s =

If you do the calculations correctly you will find that these are two data sets of 11 observations with the same means and standard deviations. The mean gives an estimate of center, while the standard deviation gives an estimate of spread. Neither measure is resistant to outliers, and they do not give an indication of the shape of the distribution. On the next page, the stemplot of Data A is provided, where the data have been rounded to the nearest 10th. The stems are ones and the leaves are 10ths. Complete the stemplot of Data B and comment on the shapes of the two distributions.

Data A

```
3 | 1
4 | 7
5 |
6 | 1
7 | 3
8 | 1178
9 | 113
```

Data B

```
 5 |
 6 |
 7 |
 8 |
 9 |
10 |
11 |
12 |
```

Exercise 2.20

KEY CONCEPTS - histograms, stemplots and boxplots, mean and standard deviation

The simplest graph to start with when you have a small data set is a stemplot. The first two stems are given below. Fill in the rest of the stems and complete the stemplot. See Exercise 1.26 in this study guide for hints on how to make a stemplot.

```
4.8 | 8
4.9 |
```

Given the distribution of measurements, are $\bar{x}$ and s good measures to describe this distribution? Find their values and give an estimate of the density of the earth based on these measures. Remember, if you are doing the calculation by hand, first find $\bar{x}$ and then calculate s in steps.

Exercise 2.29

KEY CONCEPTS - standard deviation

There are two points to remember in getting to the answer — the first is that numbers "further apart" from each other tend to have higher variability than numbers closer together. The other is that repeats are allowed. There are several choices for the answer to (a) but only one for (b).

COMPLETE SOLUTIONS

Exercise 2.7

a) The stem plot is fairly symmetric with no low or high outliers. Because of this, we would expect the values of the mean and the median to be fairly close to each other.

```
4 | 23
4 | 667899
5 | 0111112244444
5 | 555566677778
6 | 0111144
6 | 589
```

b) There are 43 presidents and the mean is $\bar{x}$ = 2357/43 = 58.814 years. The ordered ages at inauguration are

42	43	46	46	47	48	49	49	50	51	*51*	51	51
51	52	52	54	54	54	54	54	55	55	55	55	56
56	56	57	57	57	57	*58*	60	61	61	61	61	64
64	65	68	69									

The minimum and maximum are 42 and 69 respectively. When finding the median and the quartiles, concentrate on their positions, and don't pay attention to the tied values in the data. Since there are 43 presidents, the median is 55 and is the 22nd smallest value which has been underlined. There are 21 ages below and 21 above the underlined observation corresponding to the median. The first quartile is the median of the 21 observations below the median's position, which corresponds to the 11th smallest observation or 51. It is underlined and in italics above. The third quartile is the 11th smallest of the 21 observations above the median position and is 58. It is underlined and in italics as well.

c) The middle half of the ages falls between the first and third quartiles, or go from 51 to 58 years. The youngest 25% fall below the first quartile, or below 51 years. Bill Clinton at 46 was in the youngest 25%.

Exercise 2.9

Data A	$\bar{x}$ = 7.501	s = 2.031
Data B	$\bar{x}$ = 7.501	s = 2.031

From the stemplots on the next page, we see two distributions with quite different shapes, but with the same means and standard deviations. Data B seems less spread out than Data A despite the fact that the standard deviations are the same. The reason for this is that the standard deviation is not a resistant measure of spread and its value has been increased by the high outlier in Data B.

Data A

```
3 | 1
4 | 7
5 |
6 | 1
7 | 3
8 | 1178
9 | 113
```

Data B

```
 5 | 368
 6 | 69
 7 | 079
 8 | 58
 9 |
10 |
11 |
12 | 5
```

Exercise 2.20

The stemplot gives a distribution that appears fairly symmetric with one outlier of 4.88, but this is not that far from the bulk of the data. In this case, $\bar{x}$ and s should provide reasonable measures of center and spread.

```
4.8 | 8
4.9 |
5.0 | 7
5.1 | 0
5.2 | 6799
5.3 | 04469
5.4 | 2467
5.5 | 03578
5.6 | 12358
5.7 | 59
5.8 | 5
```

In practice, you will be using software or your calculator to obtain the mean and standard deviation from keyed-in data. However, we illustrate the step-by-step calculations to help you understand how the standard deviation works. Be careful not to round off the numbers until the last step, as this can sometimes introduce fairly large errors when computing s.

Observation	Difference	Difference squared
x_i	$x_i - \bar{x}$	$(x_i - \bar{x})^2$
5.50	0.052100	0.002714
5.61	0.162100	0.026277
4.88	-0.567900	0.322510
5.07	-0.377900	0.142808
5.26	-0.187900	0.035306
5.55	0.102100	0.010424
5.36	-0.087900	0.007726
5.29	-0.157900	0.024932
5.58	0.132100	0.017450
5.65	0.202100	0.040845
5.57	0.122100	0.014908
5.53	0.082100	0.006740
5.62	0.172100	0.029618
5.29	-0.157900	0.024932
5.44	-0.007900	0.000062

	Observation	Difference	Difference squared (cont.)
	x_i	$x_i - \bar{x}$	$(x_i - \bar{x})^2$
	5.34	-0.107900	0.011642
	5.79	0.342100	0.117033
	5.10	-0.347900	0.121034
	5.27	-0.177900	0.031648
	5.39	-0.057900	0.003352
	5.42	-0.027900	0.000778
	5.47	0.022100	0.000488
	5.63	0.182100	0.033161
	5.34	-0.107900	0.011642
	5.46	0.012100	0.000146
	5.30	-0.147900	0.021874
	5.75	0.302100	0.091265
	5.68	0.232100	0.053870
	5.85	0.402100	0.161684
Column sums =	157.99		1.366869

The mean is 157.99/29 = 5.4479. This has been subtracted from each observation to give the second column of deviations or differences $x_i - \bar{x}$. The second column is squared to give the squared deviations or $(x_i - \bar{x})^2$ in the third column. The variance is the sum of these squared deviations divided by one less than the number of observations:

$$s^2 = \frac{1.366869}{29-1} = 0.0488168 \text{ and } s = \sqrt{0.0488168} = 0.22095$$

Use the mean of 5.4479 as the estimate of the density of the earth based on these measurements, since the distribution is approximately symmetric, and without outliers.

Exercise 2.29

a) The standard deviation is always greater than or equal to zero. The only way it can equal zero is if all the numbers in the data set are the same. Since repeats are allowed, just choose all four numbers the same to make the standard deviation equal to zero. Examples are 1, 1, 1, 1 or 2, 2, 2, 2.

b) To make the standard deviation large, numbers at the extremes should be selected. So you want to put the four numbers at zero or ten. The correct answer is 0, 0, 10, 10. You might have thought 0, 0, 0, 10 or 0, 10, 10, 10 would be just as good, but a computation of the standard deviation of these choices shows that two at either end is the best choice.

c) There are many choices for (a) but only one for (b).

CHAPTER 3

THE NORMAL DISTRIBUTIONS

OVERVIEW

This chapter considers the use of **mathematical models** (mathematical formulas) to describe the overall pattern of a distribution. The name given to a mathematical model that summarizes the shape of a histogram is a **density curve**. The density curve is an idealized histogram. The area under a density curve between two numbers represents the proportion of the data that lie between these two numbers. Like a histogram, it can be described by measures of center such as the **median** (a point such that half the area under the density curve is to the left of the point) and the **mean** (the center of gravity or balance point of the density curve). In the last section we called the mean $\bar{x}$. This is how to refer to the mean of actual observations. The mean of a density curve is referred to as μ. Likewise, the standard deviation of a density curve also has a new notation. It is referred to as σ.

One of the most commonly used density curves in statistics is the **normal curve**. Normal curves are symmetric and bell-shaped. The peak of the curve is located above the mean and median, which are equal since the density curve is symmetric. The standard deviation measures how concentrated the area is around this peak. Normal curves follow the 68-95-99.7 rule, i.e., 68% of the area under a normal curve lies within one standard deviation of the mean (illustrated in the figure below), 95% within two standard deviations of the mean, and 99.7% within three standard deviations of the mean.

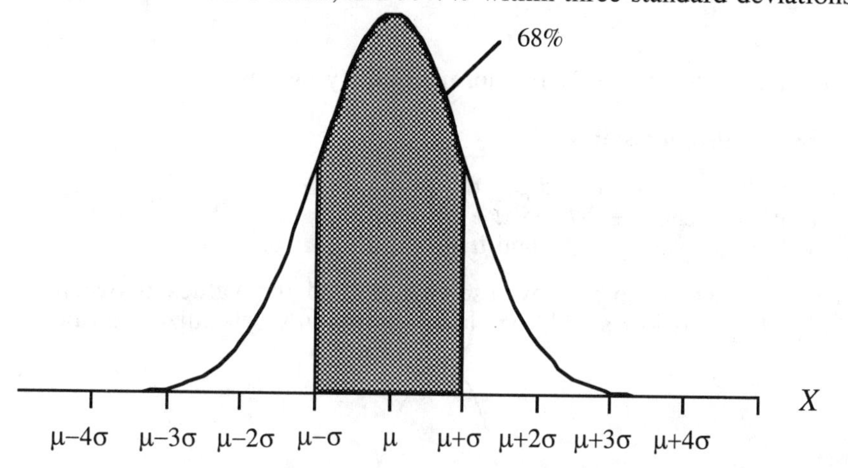

Areas under any normal curve can be found easily if quantities (x) are first standardized by subtracting the mean (μ) from each value and dividing the result by the standard deviation (σ). This **standardized value** is called the ***z*-score**.

$$z = \frac{x - \mu}{\sigma}$$

If data whose distribution can be described by a normal curve are standardized (all values replaced by their z-scores), the distribution of these standardized values is described by the **standard normal curve**. Areas under standard normal curves are easily computed by using a **standard normal table** such as that found in Table A in the front inside cover of your text.

The standard normal curve is very useful for finding the proportion of observations in an interval when dealing with any normal distribution. Here are some hints about solving these problems:

1. State the problem.
2. Draw a picture of the problem. It will help you know what area you are looking for.
3. Standardize the observations.
4. Using Table A in the front inside cover of your text, find the area you need.

HINT: The normal curve has a total area of 1. The normal curve is also symmetric so areas (proportions) such as those shown below are equal.

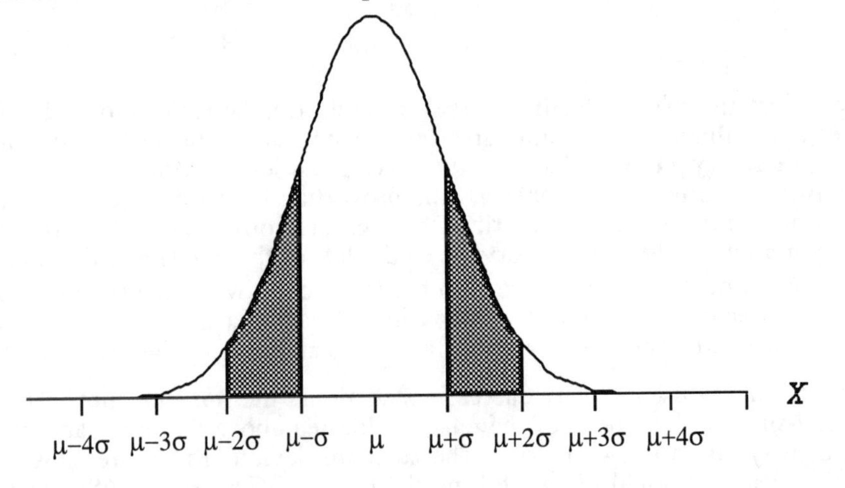

GUIDED SOLUTIONS

Exercise 3.7

KEY CONCEPTS - the 68-95-99.7 rule for normal density curves

a) Recall that the 68-95-99.7 rule states

68% of the data will be between $\mu - \sigma$ and $\mu + \sigma$,
95% of the data will be between $\mu - 2\sigma$ and $\mu + 2\sigma$, and
99.7% of the data will be between $\mu - 3\sigma$ and $\mu + 3\sigma$.

What are μ and σ in this problem? Now use this to find the values between which 95% of all pregnancies fall. The figure below should be able to help you to visualize the rule.

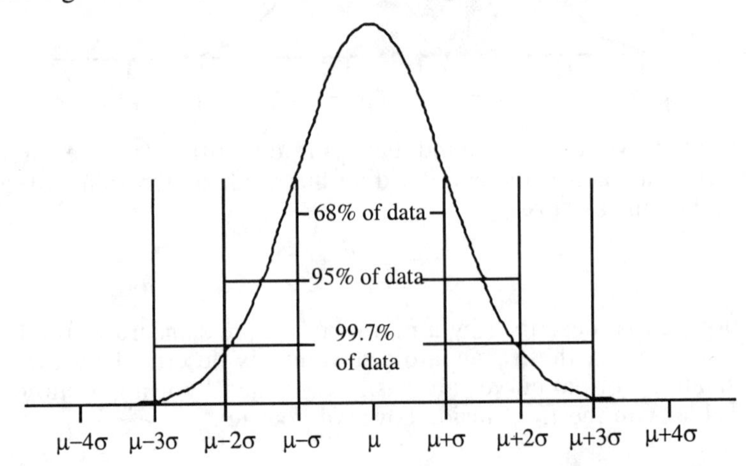

b) Refer to the figure in (a). Below what value in the figure does 2.5% of the data fall? It is one of the values $\mu - 3\sigma$, $\mu - 2\sigma$, or $\mu - \sigma$. Remember the normal curve is symmetric about μ.

Now convert this to a value in days.

Exercise 3.9

KEY CONCEPTS - standardized scores, z-scores

To compare scores from two normal distributions, each can be standardized or converted into a z-score. For example, a man's height or a woman's height that corresponds to a z-score greater than two, places either a man in the top 2.5% of men's heights or a woman in the top 2.5% of women's heights. This is because in either case the z-score corresponds to a height that is at least two standard deviations above the mean of its distribution. Using the mean and standard deviation from each distribution, convert a height of 6 feet (72 inches) tall to a z-score for men and women.

Women z-score =

Men z-score =

Which is more unusual: a 6 foot man or a 6 foot woman? The z-score helps answer this.

Exercise 3.13

KEY CONCEPTS - finding the value x (the quantile) corresponding to a given area under an arbitrary normal curve

This is an example of a "backward" normal calculation. First we *state the problem.* To make use of Table A in the front inside cover of your text, we need to state the problem in terms of areas to the left of some value. Next, we *use the table.* To do so, we think of having standardized the problem and we then find the value z in the table for the standard normal distribution that satisfies the stated condition, i.e., has the desired area to the left of it. We next must *unstandardize* this z value by multiplying by the standard deviation and then adding the mean to the result. This unstandardized value x is the desired result. We illustrate this strategy in the solutions below.

State the problem. We are told that the WISC scores are normally distributed with $\mu = 100$ and $\sigma = 15$. We want to find the score x that will place an IQ score in the lowest 25% of this distribution. This means that 25% of the population scores less than x. If the problem asked you to find the first quartile of the distribution, this would be another way of asking exactly the same question. We will first need to find the corresponding value z for the standard normal. This is illustrated in the figure below.

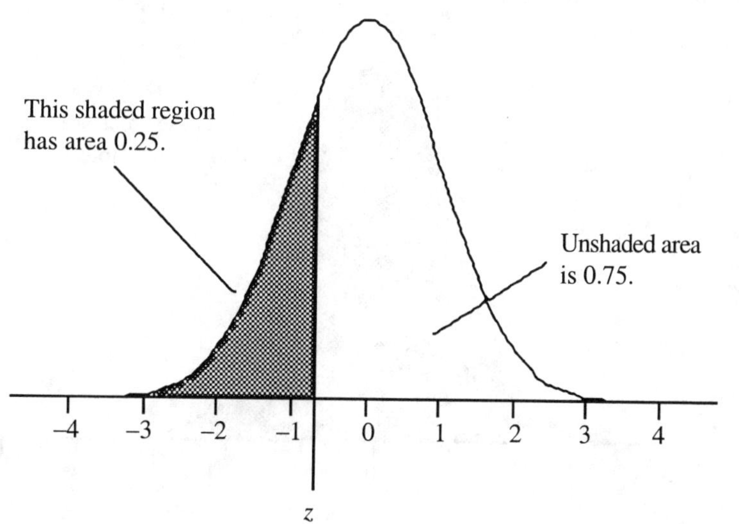

Use the table. The value z must have the property that the area to the left of it is 0.25. Areas to the left are the types of areas reported in Table A. Find the entry in the body of Table A that has a value closest to 0.25. This entry is 0.2514. The value of z that yields this area is seen, from Table A, to be –0.67.

Unstandardize. We now must unstandardize z. The unstandardized value is

$$x = (\text{standard deviation}) \times z + \text{mean} = 15z + 100 = 15 \times (-0.67) + 100 = 89.95.$$

Thus a person must score below 89.95 to be in the lowest 25%. Assuming fractional scores are not possible, a person would have to score 89 or below to score below the first quartile.

b) Now see if you can determine the score needed for a person to place in the top 5%. Use the same line of reasoning as above.

State the problem. You may find it helpful to draw the region representing the z value corresponding to the top 5%.

Use the table.

Unstandardize.

Exercise 3.18

KEY CONCEPTS - computing areas under a standard normal density curve

Recall that the proportion of observations from a standard normal distribution that are less than a given value z is equal to the area under the standard normal curve to the left of z. Table A gives these areas. This is illustrated in the figure below.

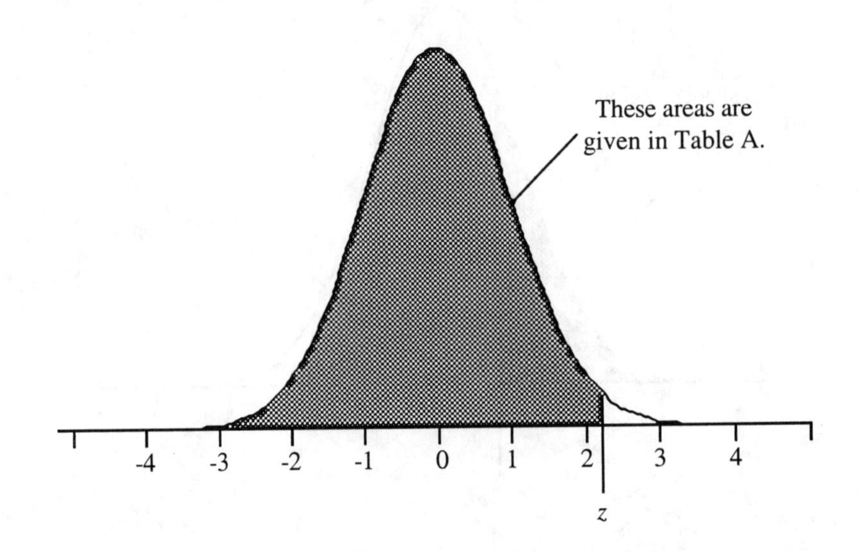

In answering questions concerning the proportion of observations from a standard normal distribution that satisfy some relation, we find it helpful to first draw a picture of the area under a normal curve corresponding to the relation. We then try to visualize this area as a combination of areas of the form in the figure on the previous page, since such areas can be found in Table A. The entries in Table A are then combined to give the area corresponding to the relation of interest.

This approach is illustrated in the solutions that follow.

a) To get you started, we will work through a complete solution. A picture of the desired area is given below.

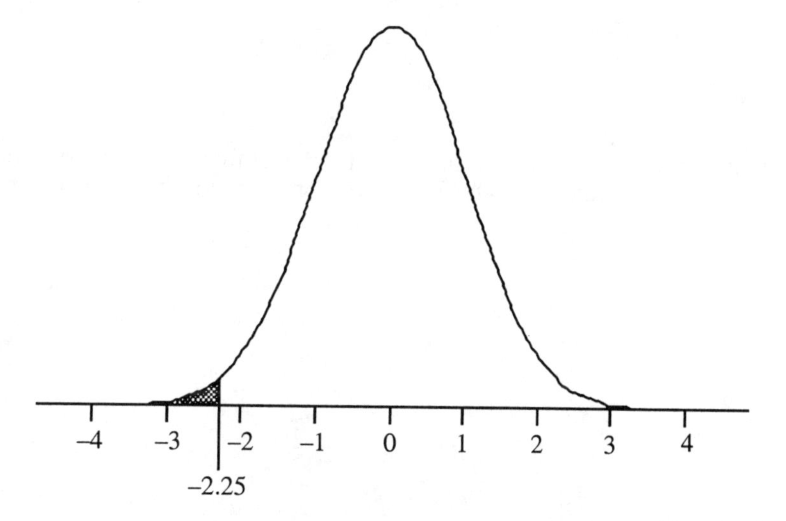

This is exactly the type of area that is given in Table A. We simply find the row labeled –2.2 along the left margin of the table, locate the column labeled .05 across the top of the table, and read the entry in the intersection of this row and column. We find this entry is 0.0122. This is the proportion of observations from a standard normal distribution that satisfies $z < -2.25$.

b) Shade the desired area in the figure below.

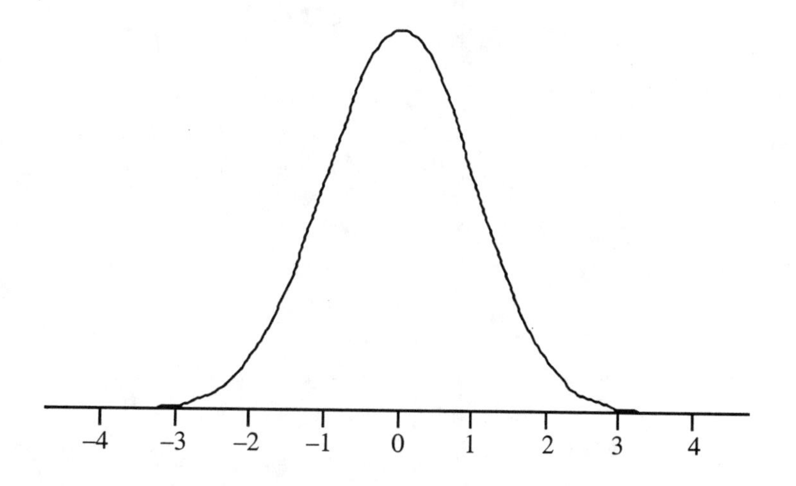

Remembering that the area under the whole curve is 1, how would you modify your answer from part (a)?

area =

c) Try solving this part on your own. To begin, draw a picture of a normal curve and shade the region.

Now use the same line of reasoning as in part (b) to determine the area of your shaded region. Remember, you want to try to visualize your shaded region as a combination of areas of the form as given in Table A.

d) To test yourself, try this part on your own. It is a bit more complicated than the previous parts, but the same approach will work. Draw a picture and then try and express the desired area as the difference of two regions for which the areas can be found directly in Table A.

COMPLETE SOLUTIONS

Exercise 3.7

a) In this problem the mean is $\mu = 266$ days and the standard deviation is $\sigma = 16$ days. From the 68-95-99.7 rule we know that the middle 95% of all pregnancies should fall between

$\mu - 2\sigma = 266 - (2 \times 16) = 266 - 32 = 234$ days and

$\mu + 2\sigma = 266 + (2 \times 16) = 266 + 32 = 298$ days.

b) Since 95% of all pregnancies fall between $\mu - 2\sigma = 234$ days and $\mu + 2\sigma = 298$ days, the remaining 5% of all pregnancies should last less than 234 days or more than 298 days. The symmetry of the normal curve about its mean implies that half of this 5% (in other words, 2.5%) will be below 234 days and the remaining 2.5% above 298 days. Thus the shortest 2.5% of all pregnancies are less than 234 days.

Exercise 3.9

Women $\quad z\text{-score} = \dfrac{72-64}{2.7} = 2.96$

Men $\quad z\text{-score} = \dfrac{72-69.3}{2.8} = 0.96$

Rounding the z-scores of 2.96 and 0.96 to 3 and 1, respectively, and then applying the 68-95-99.7 we see that about 0.15% of women are at least 6 feet tall while about 16% of men are at least 6 feet tall.

Exercise 3.13

a) The complete solution for the lowest 25% is given in the guided solutions.

b) For the top 5% we proceed as follows.

State the problem. We are told in this Exercise that the WISC scores are normally distributed with $\mu = 100$ and $\sigma = 15$. We want to find the score x that will place a person in the top 5% of this distribution. This means that 95% of the population scores less than x. We will first need to find the corresponding value z for the standard normal. This is illustrated in the figure below.

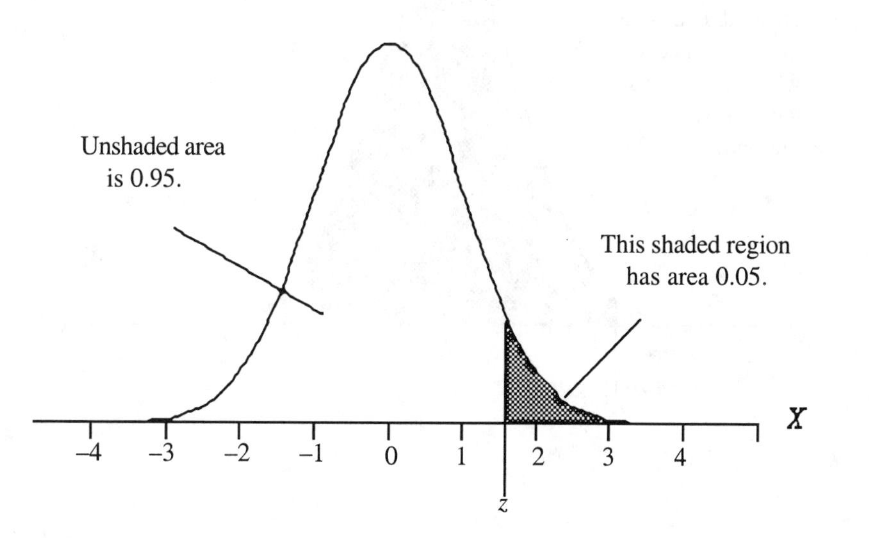

Use the table. The value z must have the property that the area to the left of it is 0.95. Areas to the left are the types of areas reported in Table A. Find the entry in the body of Table A that has a value closest to 0.95. This entry is 0.9495. The value of z that yields this area is seen, from Table A, to be 1.64.

Unstandardize. We now must unstandardize z. The unstandardized value is

$$x = (\text{standard deviation}) \times z + \text{mean} = 15z + 100 = 15 \times 1.64 + 100 = 124.6$$

Thus a person must score at least 124.6 to be in the top 5%. Assuming fractional scores are not possible, a person would have to score at least $x = 125$ to place in the top 5%.

Exercise 3.18

a) A complete solution was provided in the guided solutions.

b) The desired area is indicated below.

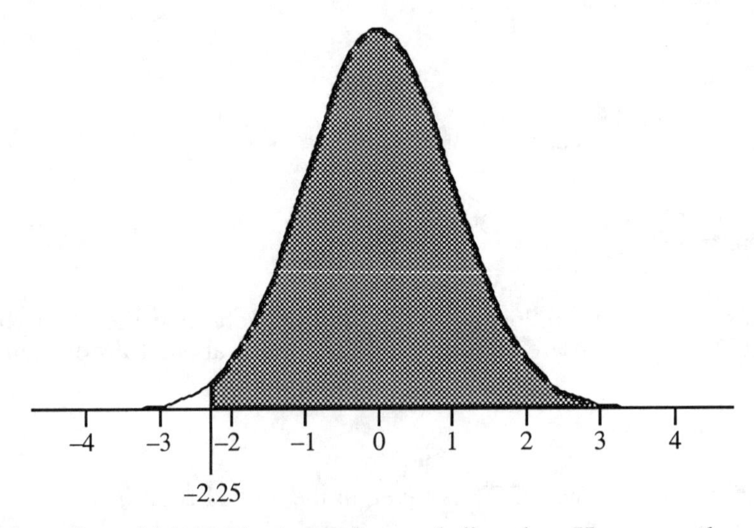
–4
–3
–2
–1
0
1
2
3
4
–2.25

This is not of the form for which Table A can be used directly. However, the unshaded area to the left of –2.25 is of the form needed for Table A. In fact, we found the area of the unshaded portion in part (a). We notice that the shaded area can be visualized as what is left after deleting the unshaded area from the total area under the normal curve.

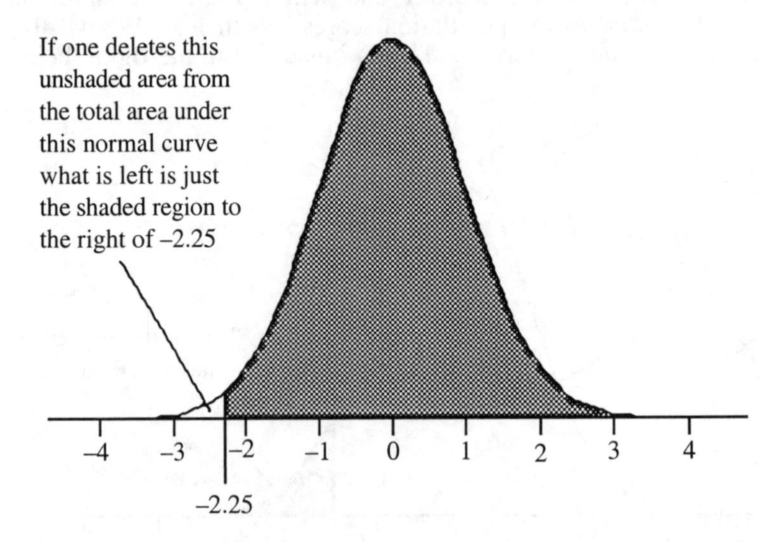
If one deletes this unshaded area from the total area under this normal curve what is left is just the shaded region to the right of –2.25
–4
–3
–2
–1
0
1
2
3
4
–2.25

Since the total area under a normal curve is 1, we have

shaded area = total area under normal curve – area of unshaded portion

= 1 – 0.0122 = 0.9878.

Thus the desired proportion is 0.9878,

c) The desired area is indicated in the figure below.

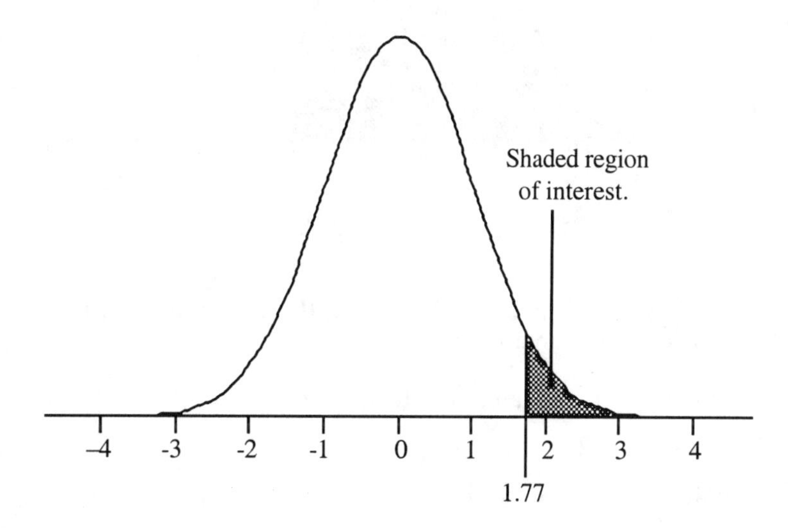

This is just like part (b). The unshaded area to the left of 1.77 can be found in Table A and is 0.9616. Thus

shaded area = total area under normal curve – area of unshaded portion

= 1 – 0.9616 = 0.0384.

This is the desired proportion.

d) We begin with a picture of the desired area.

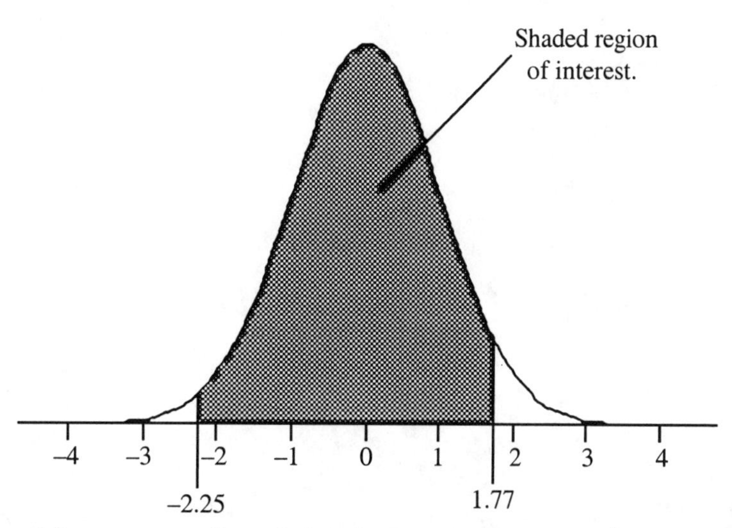

The shaded region is a bit more complicated than in the previous parts, however the same strategy still works. We note that the shaded region is obtained by removing the area to the left of –2.25 from all the area to the left of 1.77.

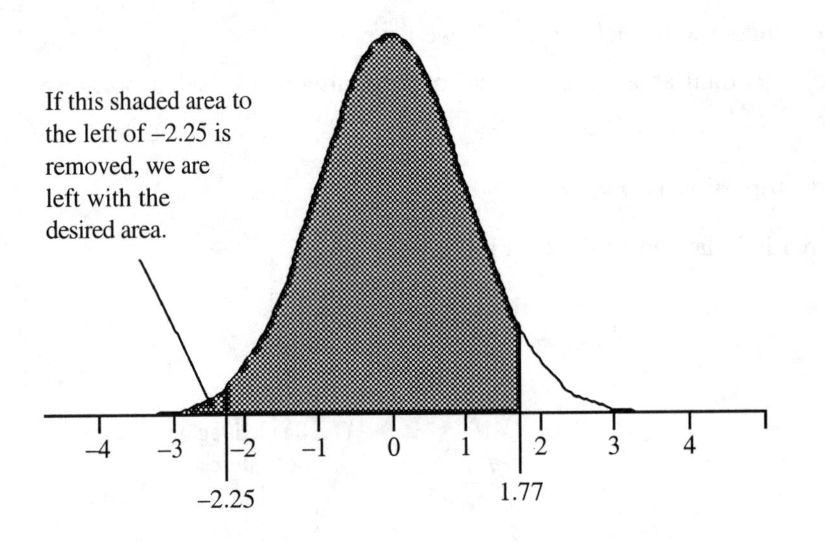
If this shaded area to the left of –2.25 is removed, we are left with the desired area.
–4
–3
–2
–1
0
1
2
3
4
–2.25
1.77

The area to the left of –2.25 is found in Table A to be 0.0122. The area to the left of 1.77 is found in Table A to be 0.9616. The shaded area is thus

$$\begin{aligned}\text{shaded area} &= \text{area to left of } 1.77 - \text{area to left of } -2.25 \\ &= 0.9616 - 0.0122 \\ &= 0.9494.\end{aligned}$$

This is the desired proportion.

CHAPTER 4

SCATTERPLOTS AND CORRELATION

OVERVIEW

Chapters 1, 2, and 3 of your textbook provide tools to explore several types of variables one by one. However, in most instances the data of interest are a collection of variables that may exhibit relationships among themselves. Typically, these relationships are more interesting than the behavior of the variables individually. In this chapter, we consider tools for exploring the relationship between variables. The first tool we consider is the **scatterplot**. Scatterplots display two quantitative variables at a time, such as the weight of a car and its miles per gallon (MPG). Using colors or different symbols, we can add information to the plot about a third variable that is categorical. For example, if in our plot we wanted to distinguish between cars with manual or automatic transmissions, we might use a circle to plot the cars with manual transmissions and a cross to plot the cars with automatic transmissions.

When drawing a scatterplot, we need to pick one variable to be on the horizontal axis and the other to be on the vertical axis. When there is a **response variable** and an **explanatory variable**, the explanatory variable is always placed on the horizontal axis. In cases where there is no explanatory-response variable distinction, either variable can go on the horizontal axis. After drawing the scatterplot by hand or using a computer, the scatterplot should be examined for an **overall pattern** that may tell us about any relationship between the variables and for **deviations** from it. You should be looking for the **direction**, **form**, and **strength** of the overall pattern. In terms of direction, **positive association** occurs when the variables both take on high values together, while **negative association** occurs if one variable takes high values when the other takes on low values. In many cases, when an association is present, the variables appear to have a form that can be described as a **linear relationship**. The plotted values seem to form a line. If the line slopes up to the right, the association is positive; if the line slopes down to the right, the association is negative. As always, look for **outliers**. The outlier may be far away in terms of the horizontal variable or the vertical variable or far away from the overall pattern of the relationship.

Scatterplots provide a visual tool for looking at the relationship between two variables. Unfortunately, our eyes are not good tools for judging the strength of the relationship. Changes in the scale or the amount of white space in the graph can easily change our judgment of the strength of the relationship. **Correlation** is a numerical measure we use to show the strength of **linear association**.

The correlation can be calculated using the formula

$$r = \frac{1}{n-1}\sum\left(\frac{x_i - \bar{x}}{s_x}\right)\left(\frac{y_i - \bar{y}}{s_y}\right)$$

where $\bar{x}$ and $\bar{y}$ are the respective means for the two variables X and Y, and s_x and s_y are their respective standard deviations. In practice, you will probably be computing the value of r using computer software or a calculator that finds r from entering the values of the x's and y's. When computing a correlation coefficient, there is no need to distinguish between the explanatory and response variables, even in cases where this distinction exists. The value of r does not change if we switch x and y.

When r is positive there is a positive linear association between the variables, and when r is negative there is a negative linear association. The value of r is always between 1 and –1. Values close to 1 or –1 show a strong association, while values near 0 show a weak association. As with means and standard deviations, the value of r is strongly affected by outliers. Their presence can make the correlation much different than it might be with the outlier removed. Finally, remember that the correlation is a measure of straight line association. There are many other types of association between two variables, but these patterns will not be captured by the correlation coefficient.

GUIDED SOLUTIONS

Exercise 4.1

KEY CONCEPTS - explanatory and response variables

a) When examining the relationship between two variables, if you hope to show that one variable can be used to explain variation in the other, remember that the response variable measures the outcome of the study, while the explanatory variable explains changes in the response variable. When you just want to explore the relationship between two variables like score on the math and verbal SAT, then the explanatory-response variable distinction is not important.

b) Explore the relationship between the variables or view one as explanatory and the other as a response?

In the latter case, explanatory variable = response variable =

c) Explore the relationship between the variables or view one as explanatory and the other as a response?

In the latter case, explanatory variable = response variable =

d) Explore the relationship between the variables or view one as explanatory and the other as a response?

In the latter case, explanatory variable = response variable =

e) Explore the relationship between the variables or view one as explanatory and the other as a response?

In the latter case, explanatory variable = response variable =

Exercise 4.6

KEY CONCEPTS - drawing and interpreting a scatterplot

a) When drawing a scatterplot, we first need to pick one variable (the explanatory variable) to be on the horizontal axis and the other (the response) to be on the vertical axis. In this data set we are interested in the "effect" of speed on fuel used. So the speed that the car is driven is the explanatory variable and fuel used is the response variable in the plot in the following figure. Although you will generally draw scatterplots on the computer, drawing a small one like this by hand makes sure that you understand what the points represent. On the next page, we have plotted the first two points corresponding to speeds of 10 and 20 km/hr. You should complete the plot by hand.

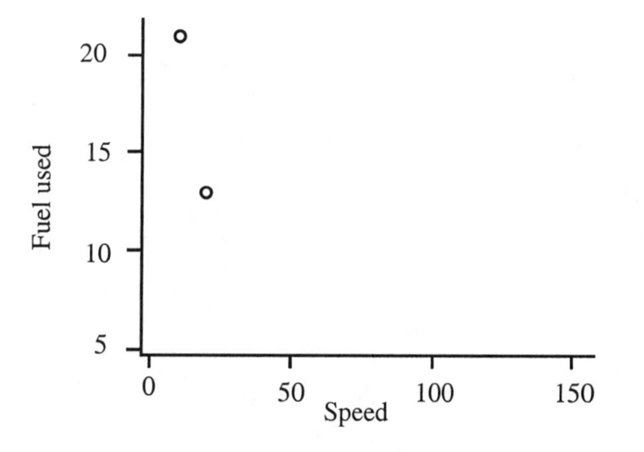

b) How would you describe the pattern in your plot in words?

Explain why this pattern makes sense. (Based on your experience, how does fuel consumption vary with speed?)

c) Positive association occurs when the variables both take on high values together, while negative association occurs if one variable takes high values when the other takes on low values. Are either of these patterns present here?

d) How close to a simple curved pattern do the points in the plot lie? If the points lie close to a simple curve, we would say the relationship is strong. If the points are scattered around a curved pattern and do not lie close to it, we would say the relationship is weak.

Exercise 4.7

KEY CONCEPTS - drawing and interpreting a scatterplot, adding a categorical variable to a scatterplot

a) When drawing a scatterplot, we first need to pick one variable (the explanatory variable) to be on the horizontal axis and the other (the response) to be on the vertical axis. In this data set we are interested in the "effect" of time on length. So time is the explanatory variable and length of the icicle is the response. To include the categorical variable, run, use a different plotting symbol for the data from runs 8903 and 8905. Try using an o for points corresponding to run 8903 and an x for points corresponding to run 8905. We have drawn the points corresponding to a time of 10 minutes for both runs in the plot on the next page. You should complete the plot, either by hand or by using software.

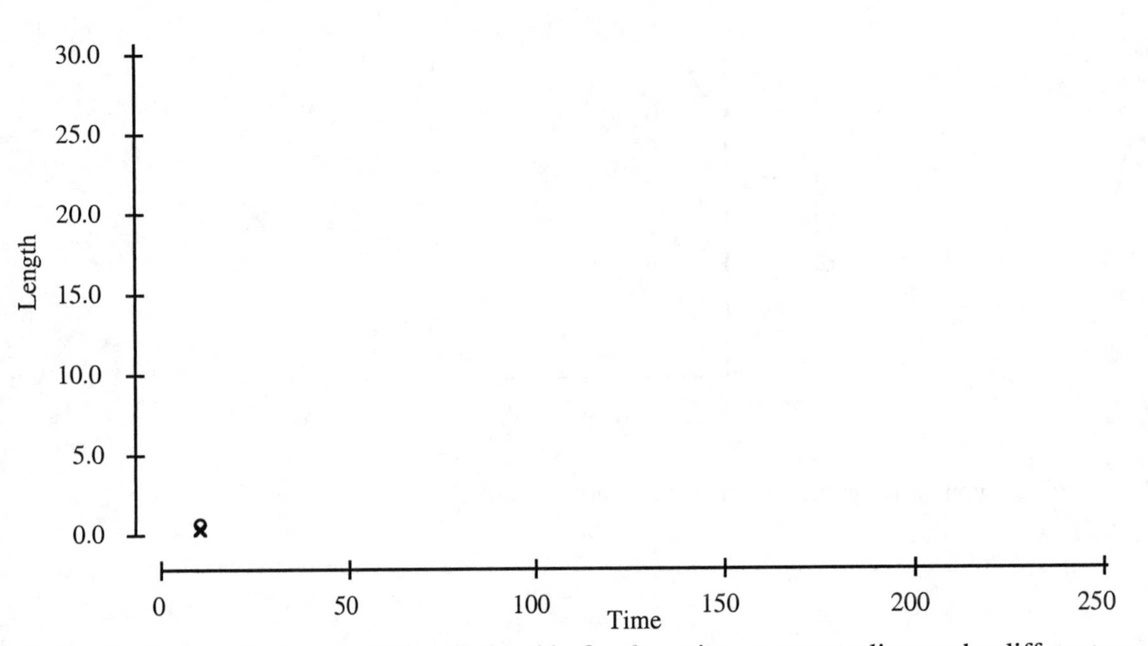

b) Describe the form and strength of the relationship for the points corresponding to the different runs. Can the relationships be described with a straight line? How do the patterns for the two runs differ?

Exercise 4.8

KEY CONCEPTS - scatterplots and computing the correlation coefficient

a) For these data we do not envision one of the variables as explaining the other. All we are interested in is investigating the association between the two variables. We are free to arbitrarily designate one of the variables as the explanatory variable and the other as the response. We choose the variable Femur as the explanatory variable, plotting it on the horizontal axis, and choose Humerus to play the role of the response, plotting it on the vertical axis. Use the axes provided to make your scatterplot.

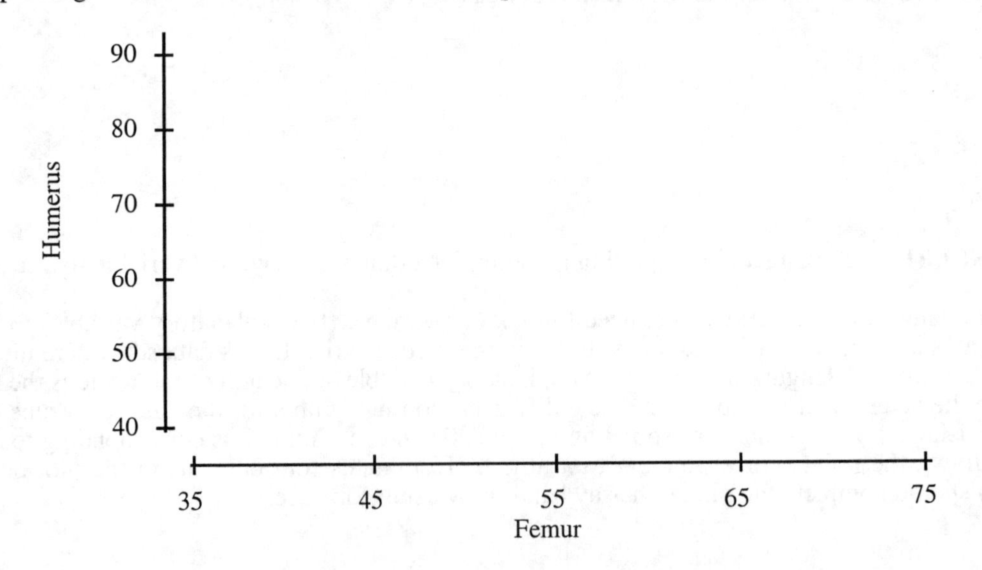

Do any of the points appear to depart from a straight line? What does this imply about whether all five specimens came from the same species?

b) Let x denote the Femur measurements and y the Humerus measurements. We find the means and standard deviations to be

$$\bar{x} = 58.20, \qquad s_x = 13.1985$$

$$\bar{y} = 66.00, \qquad s_y = 15.8902$$

Calculations by hand are best done systematically, such as in the following table. The second and fourth columns are the standardized values for x and y. We have provided the table entries for the first two x, y values. See if you can complete the remaining entries

x	$\left(\frac{x-\bar{x}}{s_x}\right)$	y	$\left(\frac{y-\bar{y}}{s_y}\right)$	$\left(\frac{x-\bar{x}}{s_x}\right)\left(\frac{y-\bar{y}}{s_y}\right)$
38	–1.5305	41	–1.5733	2.4079
56	–0.1667	63	–0.1888	0.0315
59		70		
64		72		
74		84		

Now sum up the values in the last column and divide by $n - 1$ to compute r.

$r =$

c) Use your calculator to compute r. You may need to consult your owners manual if you do not know how to compute r using your calculator.

Exercise 4.26

KEY CONCEPTS - changing units of measurement

a) You can use software to construct the plot, or prepare it by hand using the axes provided below.

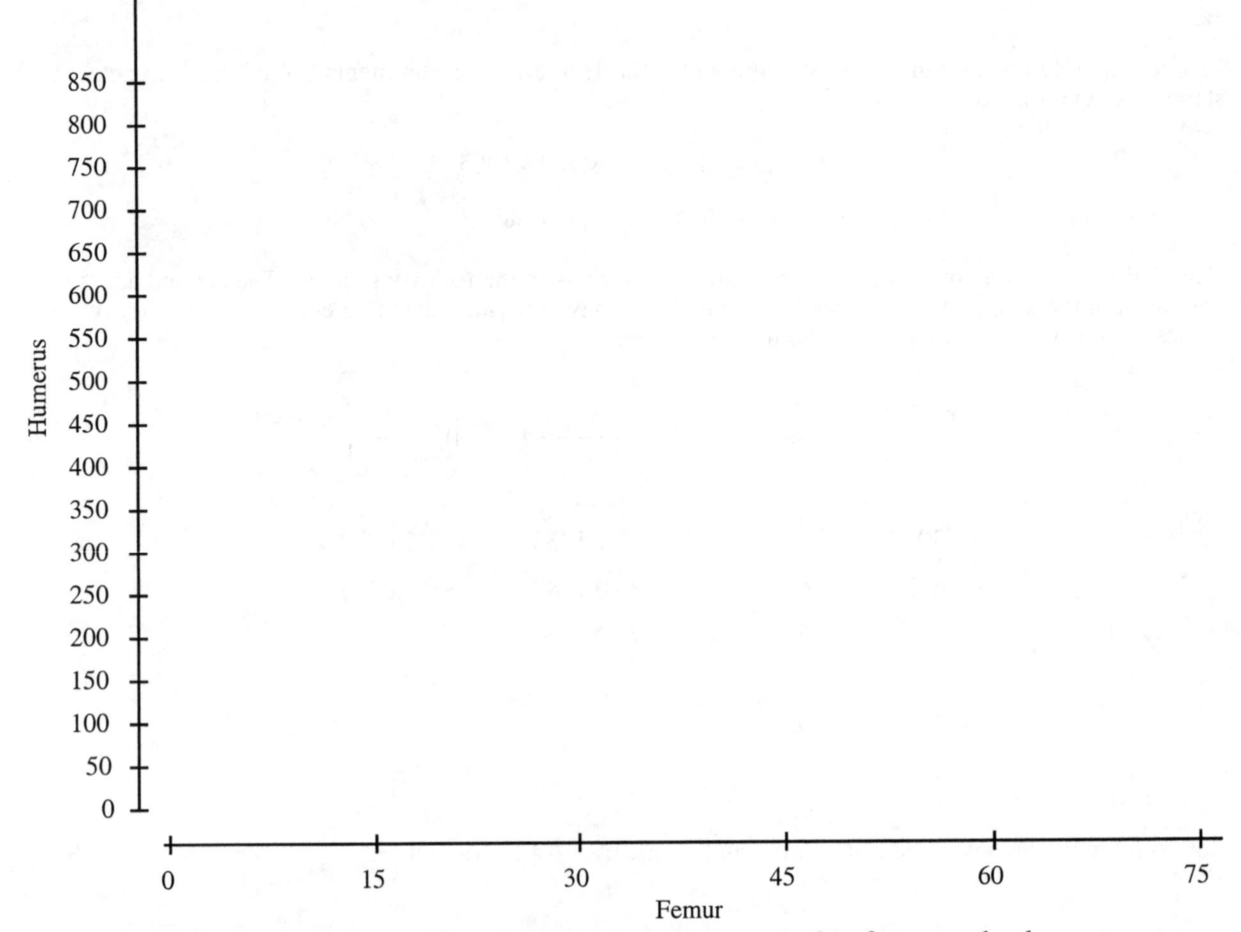

b) You may find it helpful to look at Fact 3 about correlation on page 91 of your textbook.

Exercise 4.29

KEY CONCEPTS - interpreting the correlation coefficient

a) The key phrase is "A well-diversified portfolio includes assets with low correlations." What does this imply about which of the two investments she should choose?

b) What can you say about the correlation between two variables when increases in one of the variables are associated with decreases in the other?

COMPLETE SOLUTIONS

Exercise 4.1

a) A complete solution was provided in the Guided Solutions.

b) We would probably simply want to explore the relationship between weight and height.

c) We would probably view inches of rain as explaining the yield of corn. Thus, the response is the yield of corn in bushels and the explanatory variable is inches of rain in the growing season.

d) We would probably simply want to explore the relationship between a student's scores on the SAT math exam and scores on the SAT verbal exam.

e) We would probably view a family's income as explaining the years of education their eldest child receives. Thus, the response is the years of education that the eldest child receives and the explanatory variable is the family's income.

Exercise 4.6

a) The complete scatterplot is given below.

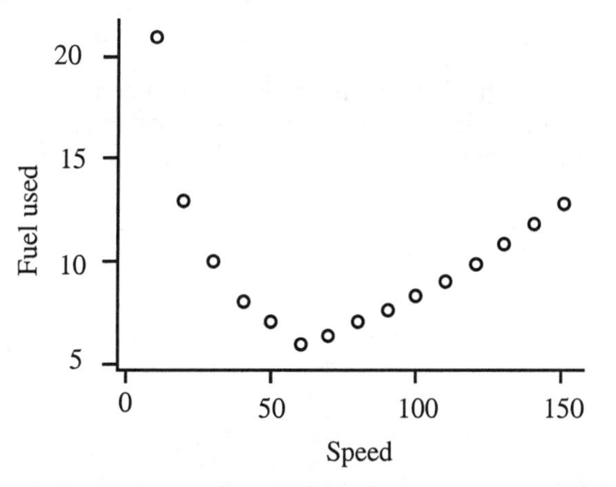

b) The plot shows a curved relationship. Fuel used first decreases as speed increases, and then at about 60 km/h increases as speed increases. This is not surprising. At very slow speeds and at very high speeds, engines are very inefficient and use more fuel, while at moderate speeds engines are more efficient and use less fuel.

c) Variable are positively associated when both take on high values together and both take on low values together. Negative association occurs when high values of one variable are associated with low values of the other. In the scatterplot, both low and high speeds correspond to high values of fuel used, so we cannot say that the variables are positively or negatively associated.

d) The points appear to lie close to a simple curved form, so we would say the relationship is reasonably strong.

Exercise 4.7

a) The scatterplot for all the data is given below.

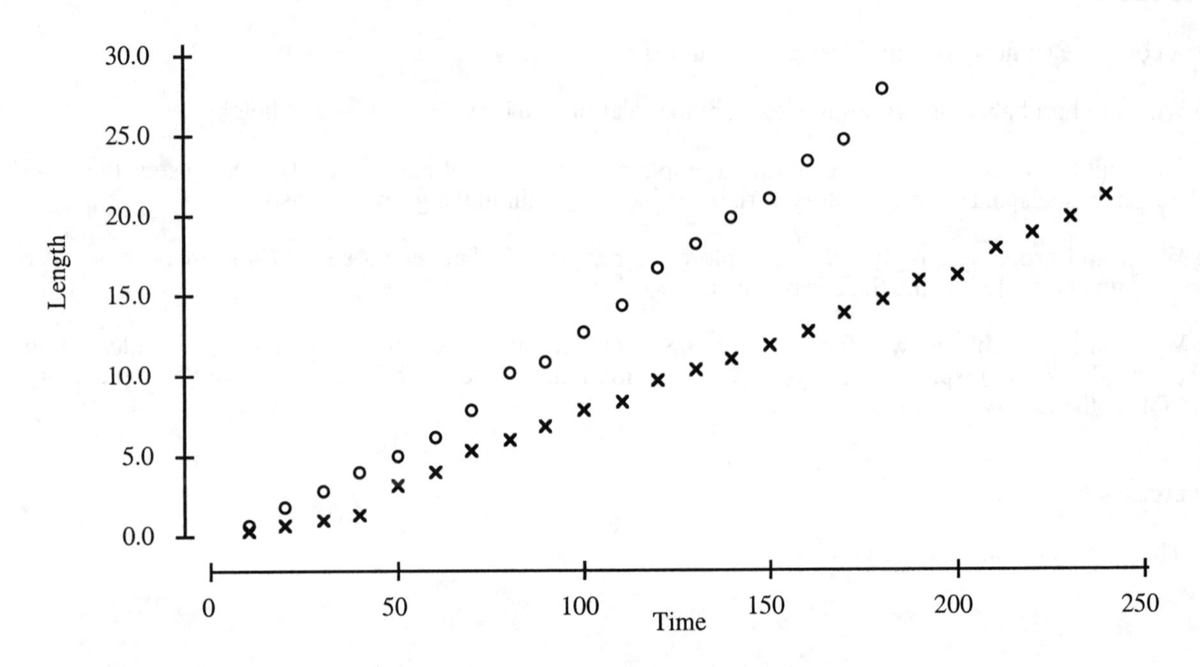

b) For both runs, there is a strong straight line relationship between time and length. Not surprisingly, length increases with time. As one might expect, the rate of growth is faster for run 8903 (the o's in the plot) that corresponds to the higher rate of water flow than for run 8905 (the x's in the plot).

Exercise 4.8

a) The scatterplot follows.

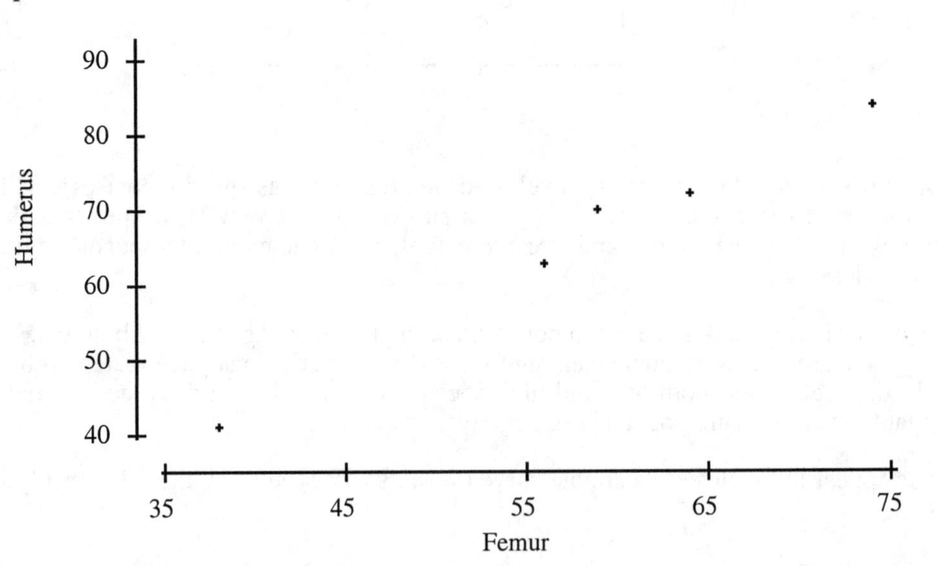

One of the points (the third point, with Femur = 59 and Humerus = 70) appears to differ a bit from the others. If this point is ignored, the other four appear to lie nearly exactly on a straight line. While the difference does not appear to be dramatic, this point might come from a different species than the others. The evidence does not appear overwhelming, however.

b) The completed table follows.

x	$\left(\frac{x - \bar{x}}{s_x}\right)$	y	$\left(\frac{y - \bar{y}}{s_y}\right)$	$\left(\frac{x - \bar{x}}{s_x}\right)\left(\frac{y - \bar{y}}{s_y}\right)$
38	–1.5305	41	–1.5733	2.4079
56	–0.1667	63	–0.1888	0.0315
59	0.0606	70	0.2517	0.0153
64	0.4394	72	0.3776	0.1659
74	1.1971	84	1.1328	1.3561

The sum of the values in the last column is 3.9767. Thus the correlation is

$$r = 3.9767/4 = 0.9942.$$

c) Our calculator gives a correlation of 0.994. This agrees with the result in part b).

Exercise 4.26

The scatterplot is given below. The points represented by the symbol + are the original data, and the points represented by the symbol x are those of the mad scientist. The patterns for the two sets of data look very different.

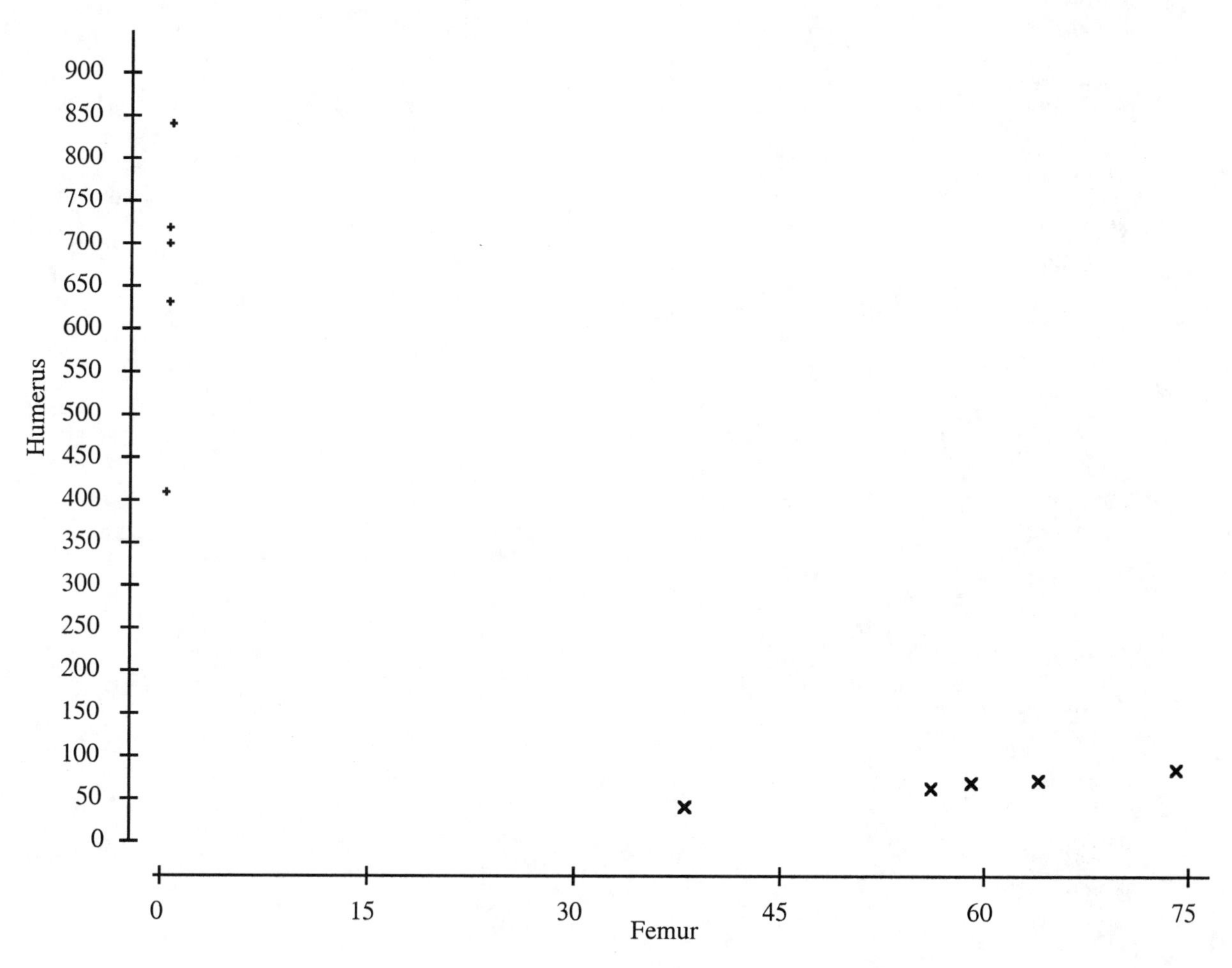

b) According to Fact 3 on page 91 of your textbook, "Because r uses the standardized values of the observations, r does not change when we change the units of measurement of x, y, or both." Thus, we know, without doing any calculations, that the correlation is exactly the same for the two sets of measurements. If you calculate the correlation for each, you find both have a correlation of 0.9942.

Exercise 4.29

a) Rachel should be looking for investments that have a weak correlation with municipal bonds, so she should invest in the small-cap stocks. The correlation between municipal bonds and the small-cap stock is weaker (closer to 0) than the correlation between municipal bonds and the large-cap stocks.

b) When increases in one variable are associated with decreases in another, the two variables are negatively associated and the correlation is negative. Thus Rachel should look for assets that are negatively correlated with municipal bonds. Neither large-cap nor small-cap stocks meet this criterion.

CHAPTER 5

REGRESSION

OVERVIEW

If a scatterplot shows a linear relationship that is moderately strong as measured by the correlation, we can draw a line on the scatterplot to summarize the relationship. In the case where there is a response and an explanatory variable, the **least-squares regression** line often provides a good summary of this relationship. A straight line relating y to x has the form

$$y = a + bx,$$

where b is the **slope** of the line and a is the **intercept**. The slope tells us the change in y corresponding to a one unit increase in x. The intercept tells us the value of y when x is 0. This has no practical meaning unless 0 is a value that x takes in practice.

The least-squares regression line is the straight line $\hat{y} = a + bx$, which minimizes the sum of the squares of the vertical distances between the line and the observed values y. The formula for the slope of the least squares line is

$$b = r\frac{s_y}{s_x}$$

and for the intercept is $a = \bar{y} - b\bar{x}$, where $\bar{x}$ and $\bar{y}$ are the means of the x and y variables, s_x and s_y are their respective standard deviations, and r is the value of the correlation coefficient. Typically, the equation of the least-squares regression line is obtained by computer software or a calculator with a regression function. The least-squares regression line can be used to predict the value of y for any value of x. Just substitute the value of x into the equation of the least-squares regression line to get the predicted value for y.

Correlation and regression are clearly related, as can be seen from the equation for the slope, b. However, the more important connection is how r^2, the square of the correlation coefficient, measures the strength of the regression. r^2 tells us the fraction of the variation in y that is explained by the regression of y on x. The closer r^2 is to 1, the better the regression describes the connection between x and y.

An examination of the **residuals** shows us how well our regression does in predictions. The difference between an observed value of y and the predicted value obtained by least-squares regression, $\hat{y}$, is called the residual.

$$\text{residual} = y - \hat{y}$$

Plotting the residuals is a good way to check the fit of our least-squares regression line. Features to look for in a **residual plot** are unusually large values of the residuals (outliers), nonlinear patterns, and uneven variation about the horizontal line through zero (corresponding to uneven variation about the regression line). Also look for influential observations. **Influential observations** are individual points whose removal would cause a substantial change in the regression line. Influential observations are often outliers in the horizontal direction.

Correlation and regression must be interpreted with caution. Watch out for the following:

- Do not **extrapolate** beyond the range of the data.

• Be aware of possible **lurking variables**.

• Are the data averages or from individuals? **Averaged data** usually lead to overestimating the correlations.

• Most of all, remember that *association is not causation!* Just because two variables are correlated doesn't mean one causes changes in the other. The best evidence that an observed association is due to causation comes from a carefully designed experiment.

GUIDED SOLUTIONS

Exercise 5.3

KEY CONCEPTS - drawing and interpreting the least-squares regression line, least-squares regression, prediction

a) Using statistical software we obtain the following output.

R squared = 96.2% R squared (adjusted) = 96.0%
s = 1.608 with 22 – 2 = 20 degrees of freedom

Source	Sum of Squares	df	Mean Square	F-ratio
Regression	1308.08	1	1308.08	506
Residual	51.7378	20	2.58689	

Variable	Coefficient	s.e. of Coeff	t-ratio	prob
Constant	9.13983	0.8208	11.1	≤ 0.0001
City	0.855972	0.0381	22.5	≤ 0.0001

The intercept can be found in the column headed "Coefficient" in the next to last line of the output (the line labeled Constant), the slope in the same column, last line (the line labeled City). After rounding, we verify that the equation of the least-squares regression line is

highway mileage = 9.140 + 0.856 (city mileage)

To make a scatterplot of the data, use statistical software or the axes provided below.

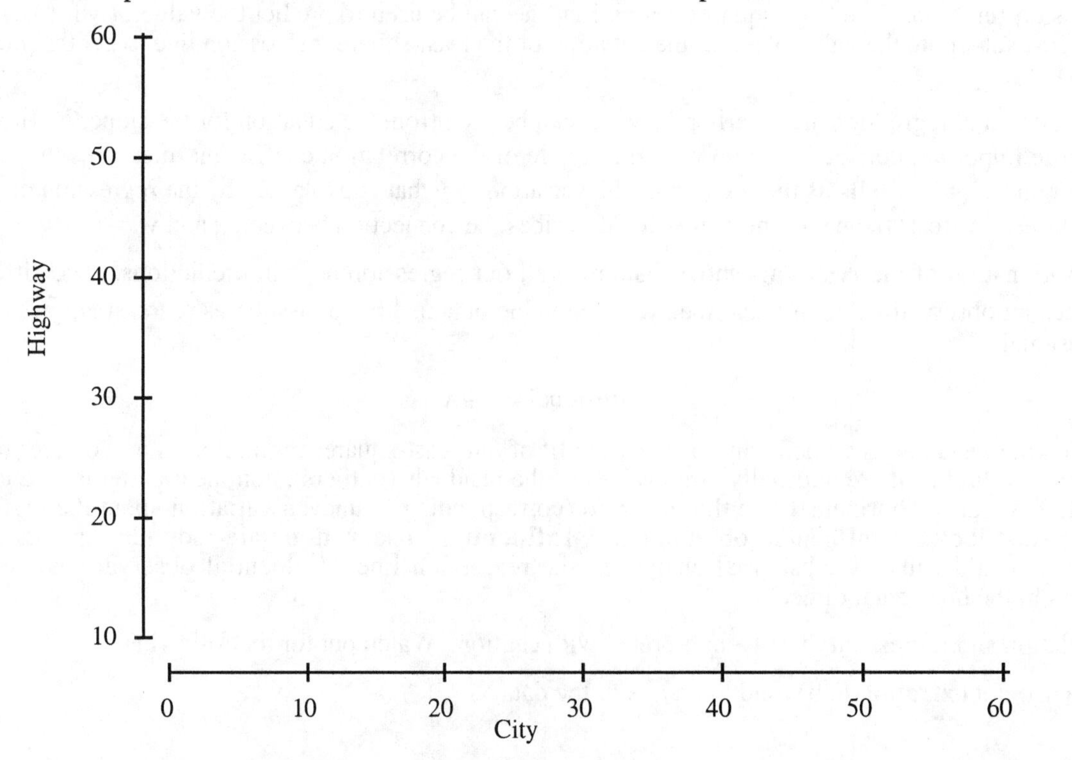

Plot the regression line on this graph. Perhaps the simplest way to draw a graph of the least-squares regression line is to pick two convenient values for the variable city mileage, substitute them into the equation of the least-squares regression line, and compute the corresponding value of highway mileage predicted by the equation for each. This produces two sets of city mileage and highway mileage values. Each of these city and highway mileage pairs corresponds to a point on the least-squares regression line. Simply plot these points on your graph and connect them with a straight line.

Convenient values for city mileage might be 0 and 50. Complete the following.

For city mileage = 0, highway mileage =

For city mileage = 50, highway mileage =

Now plot these two sets of values in the plot above. Then connect them with a straight line.

b) For a line whose equation is

$$y = a + bx$$

the quantity b is the slope. Identify this quantity in the equation of the least-squares regression line you found in part (a).

Slope =

Explain clearly in writing what this slope says about the change in the highway mileage of two-seater cars. (Writing your explanation out forces you to express your thoughts explicitly. If you can't write a clear explanation, you may not adequately understand what the slope means.)

c) To predict the highway mileage of another two-seater that is rated at 20 miles per gallon in the city, substitute city mileage = 20 into the equation of the least-squares regression line.

highway mileage = 9.140 + 0.856(city mileage) =

Exercise 5.9

KEY CONCEPTS - outliers and influential observations

a) We discussed how to prepare the requested plot in the guided solution for Exercise 5.3 previously in this study guide. Use statistical software or the axes provided below to create your plot.

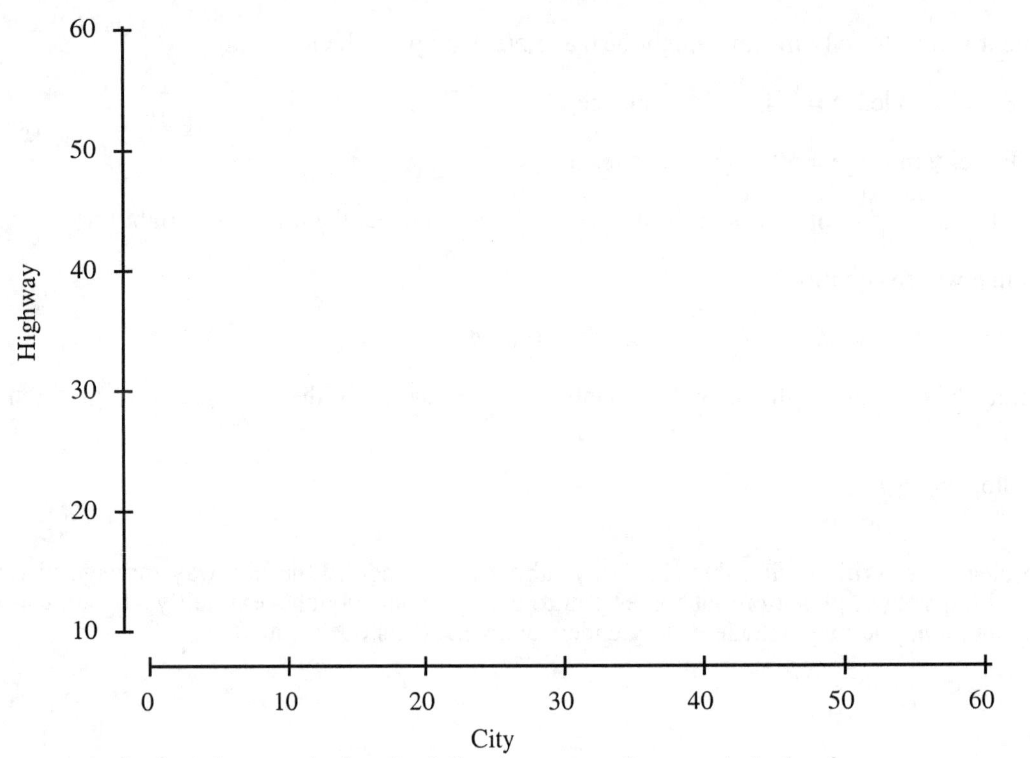

b) If one removes the Insight, one obtains the following output from statistical software.

R squared = 92.7% R squared (adjusted) = 92.3%
s = 1.255 with 21 – 2 = 19 degrees of freedom

Source	Sum of Squares	df	Mean Square	F-ratio
Regression	381.328	1	381.328	242
Residual	29.9098	19	1.57420	

Variable	Coefficient	s.e. of Coeff	t-ratio	prob
Constant	5.01163	1.280	3.91	0.0009
City	1.09293	0.0702	15.6	$\leq$ 0.0001

The intercept can be found in the column headed "Coefficient" in the next to last line of the output (the line labeled Constant), the slope in the same column, last line (the line labeled City). What do you find for the equation of the least-squares regression line ?

Add this line to your plot in part (a).

c) The equations of the two regression lines are:

with the Insight: highway mileage = 9.140 + 0.856(city mileage)

without the Insight: highway mileage = 5.012 + 1.093(city mileage)

Substitute 10, 20, and 25 MPG for city mileage into each of these equations to get the desired predictions.

predicted highway mileage when city mileage = 10:

line with Insight:

line without Insight:

predicted highway mileage when city mileage = 20:

line with Insight:

line without Insight:

predicted highway mileage when city mileage = 25:

line with Insight:

line without Insight:

Do you think the Insight changes the predictions thought to be important to a car buyer?

Exercise 5.10

KEY CONCEPTS - scatterplots, least-squares regression, r^2, extrapolation

a) On the axes provided, draw the scatterplot. Which is the response variable and which the explanatory variable? Which goes on the vertical and which on the horizontal axis? Be sure to label axes clearly.

Use your calculator or statistical software to find the least-squares regression line. Write the equation.

b) What quantity in the least-squares regression line represents the average decline per year in farm population during this period? What is its value? In answering this question, remember to keep in mind the units of the variable Population.

What quantity represents the percent of the observed variation in farm population that is accounted for by linear change over time? What is the value of this quantity?

c) Use the equation of the least-squares regression line you computed in part (a) to answer this question. Again, remember the units of the variable Population in reporting your answer. Is your prediction reasonable? Why?

Exercise 5.18

KEY CONCEPTS - calculating the least-squares regression line from summary statistics, r and r^2, prediction

a) Recall that if the least-squares regression line has the equation

$$\hat{y} = a + bx$$

the formula for the slope of the least-squares line is

$$b = r\frac{s_y}{s_x}$$

and for the intercept is

$$a = \bar{y} - b\bar{x}$$

where $\bar{x}$ and $\bar{y}$ are the means of the x and y variables, s_x and s_y are their respective standard deviations, and r is the value of the correlation coefficient.

Now use the values for $\bar{x}$, $\bar{y}$, s_x, s_y and r given in the problem to compute the slope b and intercept a.

$b =$ $a =$

Equation of least-squares regression line: $\hat{y} =$

b) What quantity tells you the fraction of the variation in the values of y (in this case GPA) that is explained by the least-squares regression of y (or GPA) on x (in this case IQ)? Compute this quantity. Remember to convert the fraction to a percent.

c) In part (a) you should have found that the equation of the least-squares regression line is

$$\text{GPA} = -3.552 + 0.101 \times \text{IQ}$$

Use this equation to predict the value of GPA for a student with an IQ of 103.

Predicted GPA =

What, therefore, is the residual for the student with an IQ of 103 and a GPA of 0.53?

Residual = observed y – predicted y =

Exercise 5.34

KEY CONCEPTS - explanatory and response variables, lurking variable

Complete the following. Think about what sort of student takes two or more years of foreign language study.

Explanatory variable =

Response variable =

Lurking variable =

Why does this lurking variable prevent the conclusion that language study improves students' English scores?

Exercise 5.41

KEY CONCEPTS - residuals, residual plots, influential observations

Recall that the least-squares regression line for predicting highway mileage from city mileage is

$$\text{highway mileage} = 9.140 + 0.856(\text{city mileage})$$

Use software or your calculator to compute the residual for each car. You can record the values in the table below.

Car Model	Residuals
Acura NSX	
Audi TT Quattro	
Audi TT Roadster	
BMW M Coupe	
BMW Z3 Coupe	
BMW Z3 Roadster	
BMW Z8	
Chevrolet Corvette	
Chrysler Prowler	
Ferrari 360 Modena	
Ford Thunderbird	
Honda Insight	
Honda S2000	
Lamborghini Murcielango	
Mazda Miata	
Mercedes-Benz SL500	
Mercedes-Benz SL600	
Mercedes-Benz SLK230	
Mercedes-Benz SLK320	
Porsche 911 GT2	
Porsche Boxster	
Toyota MR2	

If you want to make the residual plot by hand, use the axes below.

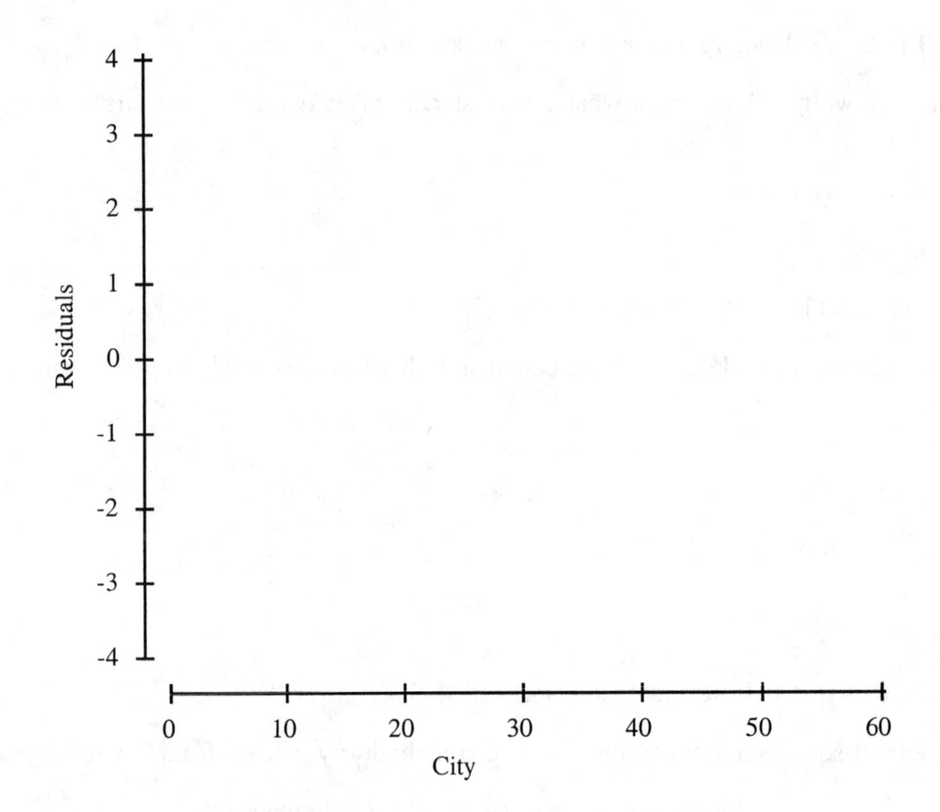

a) From your table of values of residuals,

car with largest positive residual =

car with largest negative residual =

b) The Honda Insight is an influential observation. What does this mean about how close the Honda Insight comes to falling on the least-squares line, and hence the magnitude of the residual?

c) Recall the residual = observed (or actual) y – predicted y. Use this to answer the question.

COMPLETE SOLUTIONS

Exercise 5.3

a) The equation of the least-squares regression line is given in the guided solution. A scatterplot with the regression line superimposed is given on the next page.

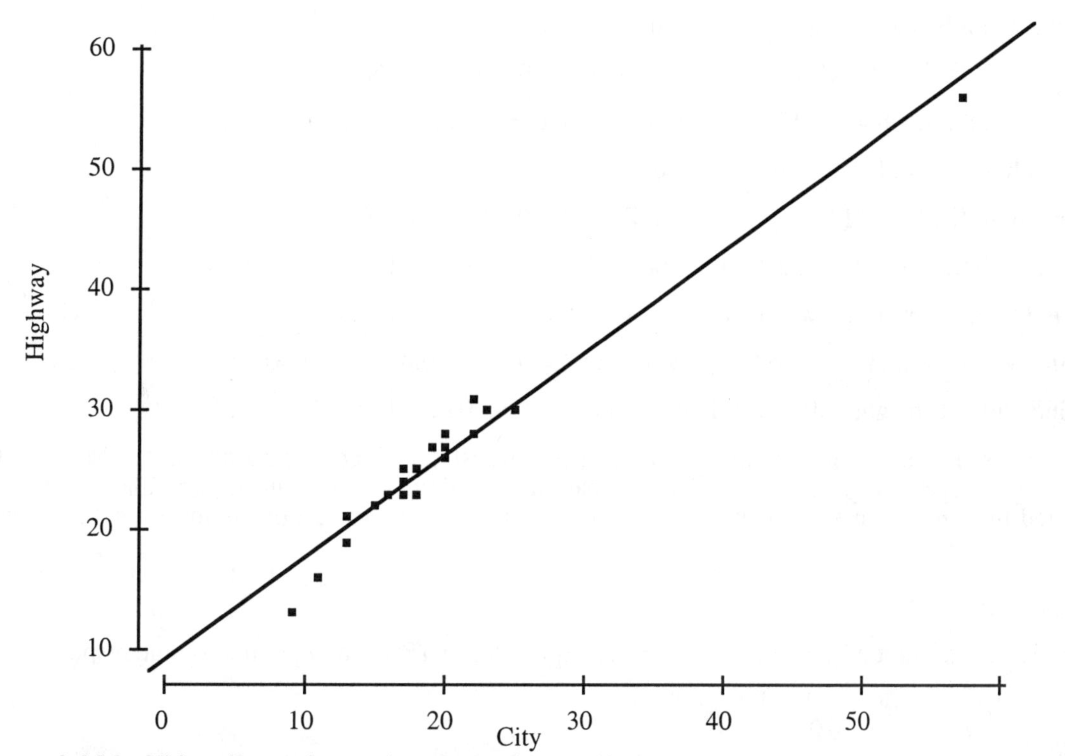

b) Slope = 0.856. This tells us that an increase of one mile per gallon in city mileage corresponds, on average, to an increase of 0.856 miles per gallon in highway mileage for two-seater cars.

c) highway mileage = 9.140 + 0.856 (city mileage) = 9.140 + 0.856(20) = 26.26

Exercise 5.9

a) and b)

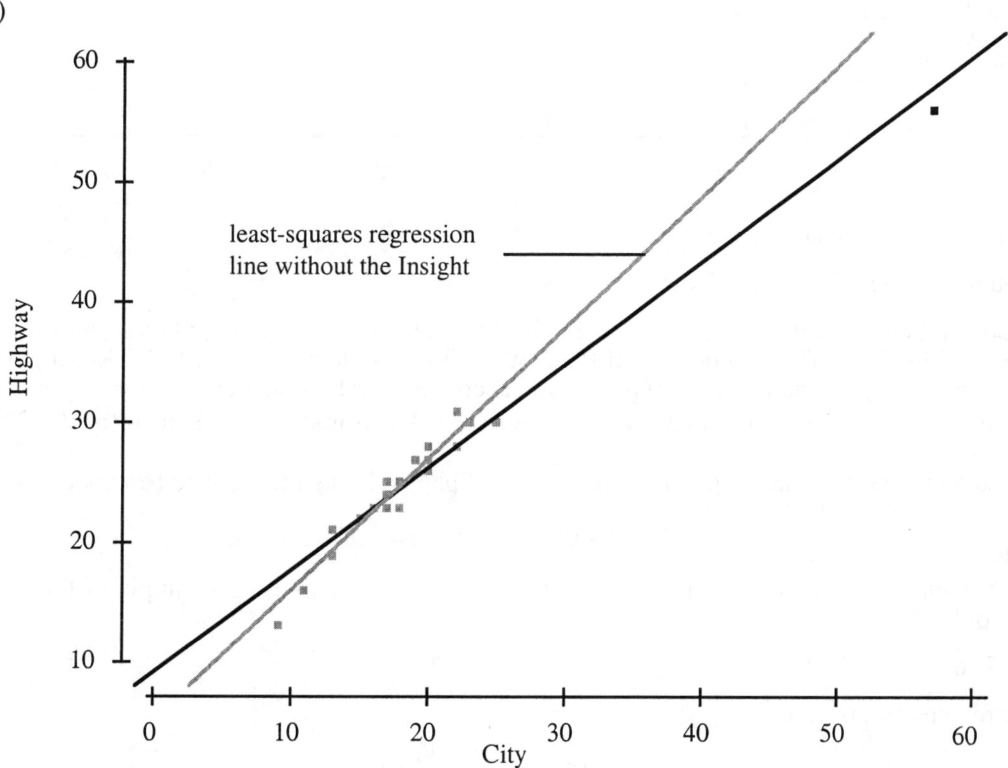

c) predicted highway mileage when city mileage = 10:

line with Insight: 9.140 + 0.856(city mileage) = 9.140 + 0.856(10) = 17.7

line without Insight: 5.012 + 1.093(city mileage) = 5.012 + 1.093(10) = 15.94

predicted highway mileage when city mileage = 20:

line with Insight: 9.140 + 0.856(city mileage) = 9.140 + 0.856(20) = 26.26

line without Insight: 5.012 + 1.093(city mileage) = 5.012 + 1.093(20) = 26.87

predicted highway mileage when city mileage = 25:

line with Insight: 9.140 + 0.856(city mileage) = 9.140 + 0.856(25) = 30.54

line without Insight: 5.012 + 1.093(city mileage) = 5.012 + 1.093(25) = 32.34

The changes in the predictions are not large (the largest difference being about 1.8 MPG). This is probably not large enough to be important to a car buyer looking at two-seaters, particularly if the buyer is interested in a sports car (in which case gas mileage is probably not one of the more important issues in choosing the car).

Exercise 5.10

a) Here is a scatterplot of the data with year the explanatory variable and population the response.

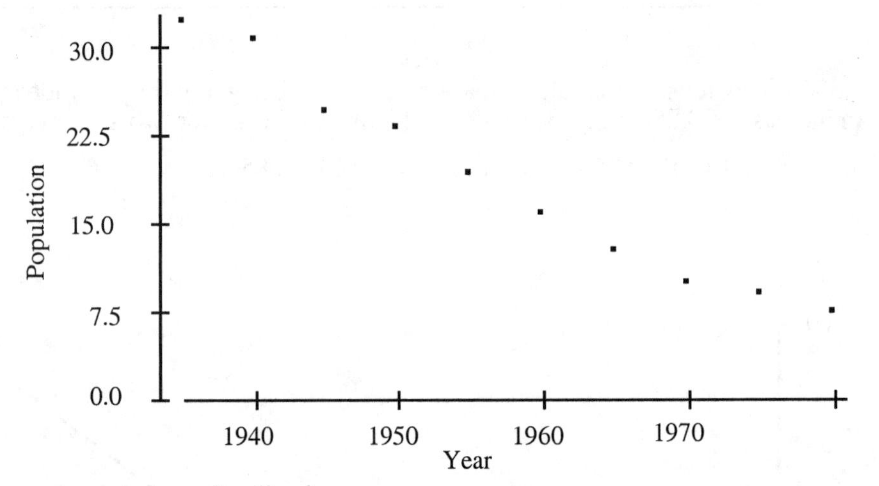

We find that the least-squares regression line is

Population = 1166.93 – 0.58679 ×Year

b) The slope of the least-squares regression line indicates the decline in farm population per year over the period represented by the data. This is a decline of 0.58679 million people per year or 586,790 people per year. The percent of the observed variation in farm population accounted for by linear change over time is determined by the value of r^2, which is 0.977 (calculated using software). The desired percent is therefore 97.7%.

c) In 1990 the regression equation predicts the number of people living on farms to be about

$$\hat{y} = 1166.93 - 0.58679(1990) = -0.782 \text{ million.}$$

This result is unreasonable since population cannot be negative. This is an example of the dangers of extrapolation!

Exercise 5.18

a) In the problem we are given that

$$\bar{x} = 108.9 \qquad s_x = 13.17$$

$$\bar{y} = 7.447 \qquad s_y = 2.10$$

$$r = 0.6337$$

Thus the slope is

$$b = r\frac{s_y}{s_x} = 0.6337\frac{2.10}{13.17} = 0.101$$

and the intercept is

$$a = \bar{y} - b\bar{x} = 7.447 - 0.101 \times 108.9 = 7.447 - 10.999 = -3.552$$

The equation of the least-squares regression line is therefore

$$\text{GPA} = -3.552 + 0.101 \times \text{IQ}$$

(Your answer may differ slightly due to rounding.)

b) Recall that the square of the correlation, r^2, is the fraction of the variation in the values of y (in this case GPA) that is explained by the least-squares regression of y on x (in this case IQ). Since here r = 0.6337,

$$r^2 = (0.6337)^2 = 0.402$$

Converting this fraction to a percent we find that the observed variation in these students' GPAs that can be explained by the linear relationship between GPA and IQ is 40.2%.

c) We use the equation of the least-squares regression line we found in part (a) namely,

$$\text{GPA} = -3.552 + 0.101 \times \text{IQ}$$

to make our prediction. We calculate

$$\text{Predicted GPA} = -3.552 + 0.101 \times 103 = 6.851$$

and hence for an observed GPA of 0.53,

$$\text{Residual} = \text{observed } y - \text{predicted } y = 0.53 - 6.851 = -6.321$$

Exercise 5.34

The explanatory variable is foreign language study. It has two values depending on whether a student has studied at least two years of a foreign language or has studied no foreign language. The response is the score on the English achievement test administered to seniors. Unfortunately for the study, students choose whether or not to take a foreign language. Academically stronger students are more likely than weaker students to study a foreign language. The academic strength of a student is the lurking variable here and prevents us from concluding that language study improves students' English scores.

Exercise 5.41

A table of the residuals and a plot of these residuals follow.

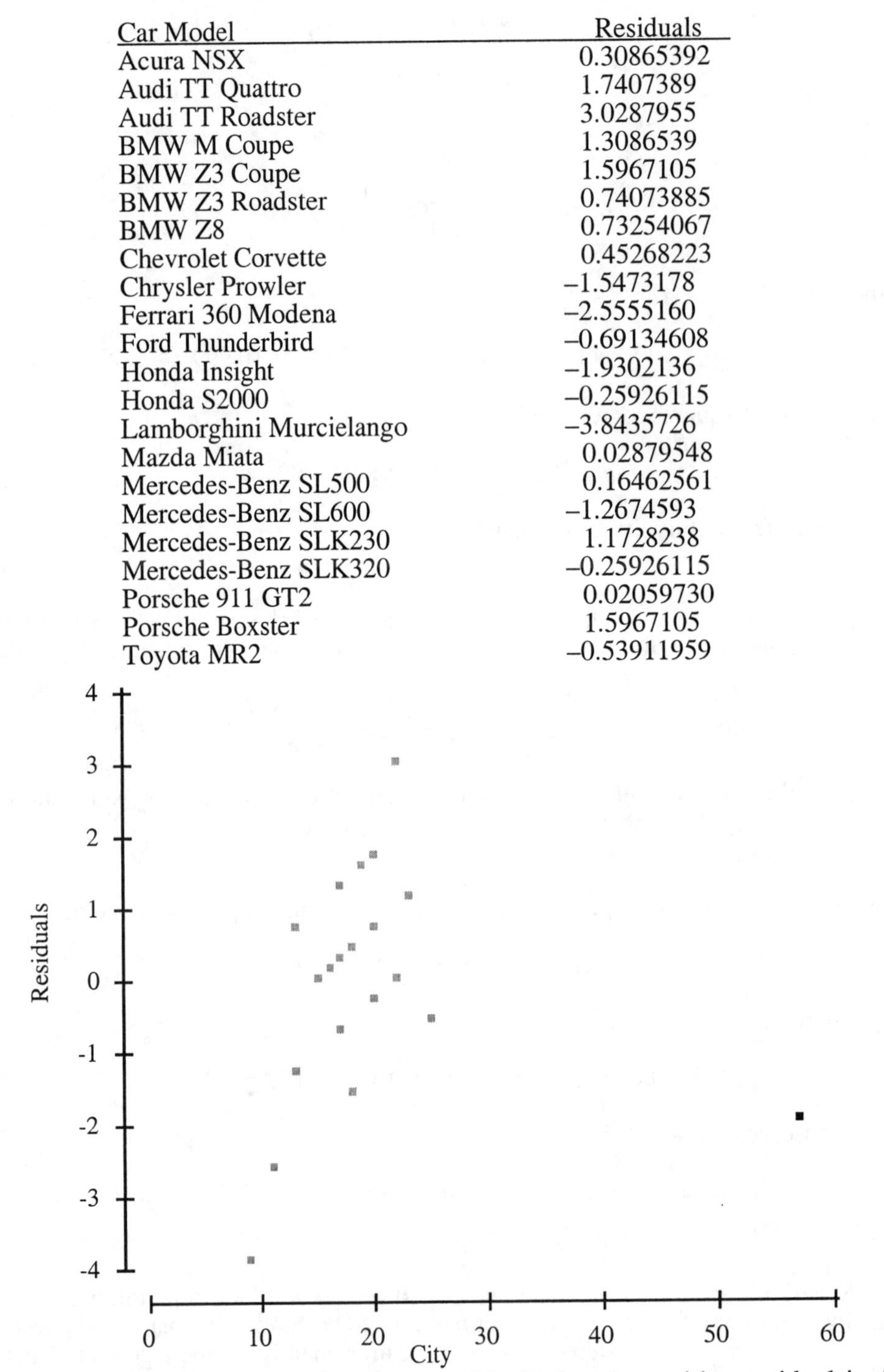

Car Model	Residuals
Acura NSX	0.30865392
Audi TT Quattro	1.7407389
Audi TT Roadster	3.0287955
BMW M Coupe	1.3086539
BMW Z3 Coupe	1.5967105
BMW Z3 Roadster	0.74073885
BMW Z8	0.73254067
Chevrolet Corvette	0.45268223
Chrysler Prowler	–1.5473178
Ferrari 360 Modena	–2.5555160
Ford Thunderbird	–0.69134608
Honda Insight	–1.9302136
Honda S2000	–0.25926115
Lamborghini Murcielango	–3.8435726
Mazda Miata	0.02879548
Mercedes-Benz SL500	0.16462561
Mercedes-Benz SL600	–1.2674593
Mercedes-Benz SLK230	1.1728238
Mercedes-Benz SLK320	–0.25926115
Porsche 911 GT2	0.02059730
Porsche Boxster	1.5967105
Toyota MR2	–0.53911959

a) From the table of residuals, we see that the car with the largest positive residual is the Audi TT Roadster. The car with the largest negative residual is the Lamborghini Murcielango.

b) The Honda Insight is an influential observation and, as a result, the least-squares regression line passes close to it. Thus, it will have a small residual.

c) The value of a residual is the actual observed value of the response minus the value predicted by the least-squares regression line. A large positive residual means that the actual highway mileage is much larger than the mileage predicted by the least-squares regression line. A large negative residual means that the actual highway mileage is much smaller than the mileage predicted by the least-squares regression line.

CHAPTER 6

TWO-WAY TABLES

OVERVIEW

This section discusses techniques for describing the relationship between two or more categorical variables. To analyze categorical variables, we use counts (frequencies) or percents (relative frequencies) of individuals that fall into various categories. **A two-way table** of such counts is used to organize data about two categorical variables. Values of the **row variable** label the rows that run across the table, and values of the **column variable** label the columns that run down the table. In each cell (intersection of a row and column) of the table, we enter the number of cases for which the row and column variables have the values (categories) corresponding to that cell.

The **row totals** and **column totals** in a **two-way table** give the marginal distributions of the two variables separately. It is usually clearest to present these distributions as percents of the table total. **Marginal distributions** do not give any information about the relationship between the variables. **Bar graphs** are a useful way of presenting these marginal distributions.

The **conditional distributions** in a two-way table help us to see relationships between two categorical variables. To find the conditional distribution of the row variable for a specific value of the column variable, look only at that one column in the table. Express each entry in the column as a percent of the column total. There is a conditional distribution of the row variable for each column in the table. Comparing these conditional distributions is one way to describe the association between the row and column variables, particularly if the column variable is the explanatory variable. When the row variable is explanatory, find the conditional distribution of the column variable for each row and compare these distributions. Side-by-side bar graphs of the conditional distributions of the row or column variable can be used to compare these distributions and describe any association that may be present.

Data on three categorical variables can be presented as separate two-way tables for each value of the third variable. An association between two variables that holds for each level of this third variable can be changed, even reversed, when the data are combined by summing over all values of the third variable. **Simpson's paradox** refers to such reversals of an association.

GUIDED SOLUTIONS

Exercise 6.3

KEY CONCEPTS - marginal distribution

a) Each of the six entries in the table correspond to a different group of people. The six different groups characterize all the people that these data describe. Add up the six entries to determine how many people these data describe.

b) The three entries in the row labeled "Arthritis" correspond to the people with arthritis of the hip or knee. Add up these three entries to get the total requested.

c) First compute the counts for each level of participation in soccer (the columns) by adding the two entries in each column. Enter these in the table below in the row labeled "Counts." The marginal distribution, as a percent, can be found from the counts in each group (elite, non-elite, did not play) that you entered in the table below. Each count must be divided by the total number of persons represented by the table that you computed in part (a). Now do the actual calculations, completing the following table.

	Group		
	Elite	Non-elite	Did not play
Counts			
Percent			

Exercise 6.5

KEY CONCEPTS - marginal distribution, describing relationships

First compute the total number of people in each group. To do this, add up the two entries in each column of the table.

Total elite =

Total non-elite =

Total did not play =

To compute the percent of each group who have arthritis, divide each entry in the row labeled "Arthritis" by the total number in the group and convert to a percent. We have done the calculation for the elite players. Fill in the remainder of the table to complete the problem.

	Group		
	Elite	Non-elite	Did not play
Percent	$10/71 \times 100\% = 14.08\%$		

Describe what these results say about the association between playing soccer and later arthritis.

Exercise 6.19

KEY CONCEPTS - marginal and conditional distributions, describing relationships

a) To compare the effectiveness of the three treatments in preventing relapse, compute the conditional distributions of relapse/no relapse for the three drugs. We have computed the percents for desipramine in the table below. You should complete the table by computing the percents for lithium and placebo.

	Desipramine	Lithium	Placebo
Relapse	$(10/24) \times 100\% = 41.67\%$		
No relapse	$(14/24) \times 100\% = 58.33\%$		

After completing the table, plot the percentages using a bar graph. Use the axes on the next page for your plot. For each of relapse and no relapse, you should plot three bars.

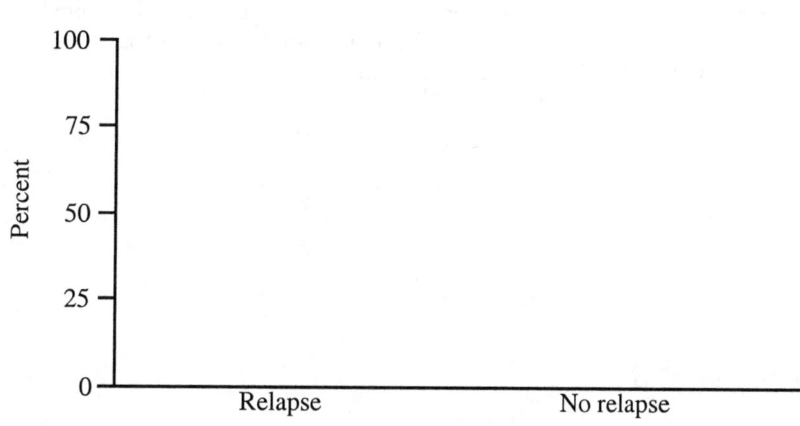

What do you observe?

b) What types of studies provide the strongest evidence for causation? What things must one be concerned about when deciding whether an observed association is due to cause-and-effect?

Exercise 6.23

KEY CONCEPTS - two-way tables, Simpson's paradox

a) Add corresponding entries in the two tables and enter the sums in the table.

	Admit	Deny
Male		
Female		

b) Convert your table in part (a) to one involving percentages of the row totals.

	Admit	Deny
Male		
Female		

c) Now repeat the type of calculations you did in part (b) for each of the original tables.

Business

	Admit	Deny
Male		
Female		

Law

	Admit	Deny
Male		
Female		

d) To explain the apparent contradiction observed in part (c), consider which professional school is easier to get into and which professional school males and females tend to apply to. Write your answer in plain English in the space provided. Avoid jargon and be clear!

COMPLETE SOLUTIONS

Exercise 6.3

a) The number of people these data describe is 10 + 9 + 24 + 61 + 206 + 548 = 858.

b) The number of people with arthritis of the hip or knee is 10 + 9 + 24 = 43.

c) The completed table is given below.

	Group		
	Elite	Non-elite	Did not play
Counts	10 + 61 = 71	9 + 206 = 215	24 + 548 = 572
Percent	71/858 × 100% = 8.28%	215/858 × 100% = 25.06%	572/858 × 100% = 66.67%

Exercise 6.5

Total elite = 10 + 61 = 71

Total non-elite = 9 + 206 = 215

Total did not play = 24 + 548 = 572

The percent of each group who have arthritis is given in the table below.

	Group		
	Elite	Non-elite	Did not play
Percent	10/71 × 100% = 14.08%	9/215 × 100% = 4.19%	24/572 × 100% = 4.20%

The percentage of cases of arthritis is higher (more than three times as high) for elite players than for non-elite players and those who did not play soccer. Non-elite players and those who did not play have nearly the same percentage of cases of arthritis. This suggests that playing lots of soccer (as represented by elite players) is associated with later arthritis.

Exercise 6.19

a) The conditional distributions of relapse/no relapse for the three drugs are given below.

	Desipramine	Lithium	Placebo
Relapse	(10/24) × 100% = 41.67%	(18/24) × 100% = 75%	(20/24) × 100% = 83.33%
No relapse	(14/24) × 100% = 58.33%	(6/24) × 100% = 25%	(4/24) × 100% = 16.67%

A bar graph that displays this information is given on the next page.

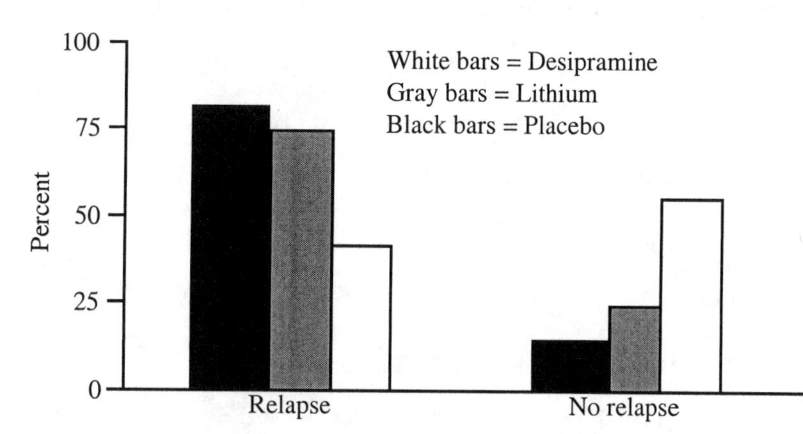

The data show that those taking desipramine had fewer relapses than those taking either lithium or a placebo.

b) These results are interesting but association does not imply causation. Some concerns are the following. First, how strong is the association? Is the number of people in the study large enough that we can conclude the observed association is strong? Second, can we rule out lurking variables? The subjects were assigned at random to the treatments, which helps guarantee that the subjects in each group are reasonably similar. However, did the subjects know anything about the treatments they received and would this knowledge affect their response to the treatment? Third, have similar results been observed in other studies with other groups in other contexts? Without additional information, one should be cautious about concluding that the study demonstrates that desipramine causes a reduction in relapses.

Exercise 6.23

a) Here is the desired two-way table.

	Admit	Deny
Male	490	210
Female	280	220

b) We first add a column containing the row totals to the table in part (a).

	Admit	Deny	Total
Male	490	210	700
Female	280	220	500

We now convert the table entries to percents of the row totals. We divide the entries in the first row by 700 and express the results as percents. We divide the entries in the second row by 500 and express these as percents.

	Admit	Deny
Male	70%	30%
Female	56%	44%

We see that Wabash admits a higher percent of male applicants.

c) We repeat the calculations in part (b), but for each of the original two tables.

Business

	Admit	Deny	Total
Male	480	120	600
Female	180	20	200

Law

	Admit	Deny	Total
Male	10	90	100
Female	100	200	300

Converting entries to percents of the row totals yields

Business

	Admit	Deny
Male	80%	20%
Female	90%	10%

	Admit	Deny
Male	10%	90%
Female	33.3%	66.7%

We see that each school admits a higher percentage of female applicants.

d) Although both schools admit a higher percentage of female applicants, the admission rates are quite different. Business admits a high percentage of all applicants; it is easier to get into the business school. Law admits a lower percentage of applicants; it is harder to get into the law school. Most of the male applicants to Wabash apply to the business school with its easy admission standards. Thus, overall, a high percentage of males are admitted to Wabash. The majority of female applicants apply to the law school. Because it has tougher admission standards, this makes the overall admission rate of females appear low, even though more females are admitted to both schools!

CHAPTER 7

PRODUCING DATA: SAMPLING

OVERVIEW

Data can be produced in a variety of ways. **Sampling**, when done properly, can yield reliable information about a **population**. **Observational studies** are investigations in which one simply observes the state of some population, usually with data collected by sampling. Even with proper sampling, data from observational studies are generally not appropriate for investigating cause-and-effect relations between variables. This is because the explanatory variable can be **confounded** with lurking variables so its effects on the response cannot be distinguished from those of the lurking variables. **Experiments** are investigations in which data are generated by active imposition of some treatment on the subjects of the experiment. Properly designed experiments are the best way to investigate cause-and-effect relations between variables.

The **population** is the entire group of individuals or objects about which we want information. The information collected is contained in a **sample** which is the part of the population we actually observe. How the sample is chosen (the sampling **design**), has a large impact on the usefulness of the data. A useful sample will be representative of the **population** and will help answer our questions. "Good" methods of collecting a sample include the following:

Simple random samples, sometime denoted **SRS**
Probability samples
Stratified random samples
Multistage samples

All these sampling methods involve some aspect of randomness through the use of a formal chance mechanism. Random selection is just one precaution that a person can take to reduce **bias,** the systematic favoring of a certain outcome. Samples we select using our own judgment, because they are convenient, or "without forethought" (mistaking this for randomness) are usually biased. This is why we use computers or a **table of random digits** to help us select a sample.

There are many ways to choose a simple random sample. A simple random sample, or SRS, of size n is a collection of n individuals chosen from the population in a manner so that each possible set of n individuals has an equal chance of being selected. In practice, simple methods such as drawing names from a hat is one way of getting an SRS. The "names in the hat" are the units in the population. To choose the sample, we mix up the "names" and select the sample of n "at random" from the hat. In reality the population may be very large. A computer or a table of random numbers can be used to "mimic" the process of "pulling the names from the hat."

The method of selecting an SRS using a table of random digits can be summed up in two steps.

1. Give every individual in the population its own numerical label. All labels need to have the same number of digits.
2. Starting anywhere in the table (usually a spot selected at random), read off labels until you have selected as many labels as needed for the sample.

Another common type of sample design is a stratified random sample. Here the population is first divided into **strata** and then an SRS is chosen from each strata. The strata are formed using some known characteristic of each individual thought to be associated with the response to be measured. Examples of strata are gender or age. Individuals in a particular stratum should be more like one another than those in the other strata.

Poor sample designs include the **voluntary response sample**, where people place themselves in the sample, and the **convenience sample**. Both of these methods rely on personal decision for the selection of the sample; this is generally a guarantee of bias in the selection of the sample.

Other kinds of bias can occur even in well designed studies. Be on the lookout for

- **Nonresponse bias** which occurs when individuals who are selected do not participate or cannot be contacted,
- Bias in the **wording of questions** leading the answers in a certain direction,
- **Confounding** or confusing the effect of two or more variables,
- **Undercoverage** which occurs when some group in the population is given either no chance or a much smaller chance than other groups to be in the sample, and
- **Response bias** which occurs when individuals do participate but are not responding truthfully or accurately due to the way the question is worded, the presence of an observer, fear of a negative reaction from the interviewer, or any other such source.

These types of bias can occur even in a randomly chosen sample and we need to try to reduce their impact as much as possible.

GUIDED SOLUTIONS

Exercise 7.1

KEY CONCEPTS - explanatory and response variables, experiments and observational studies

What are the researchers trying to demonstrate with this study? What groups are being compared and did the experiment deliberately impose membership in the groups on the subjects to observe their responses? What are the explanatory and response variables?

Explanatory variable____

Response variable____

Observational study or experiment (circle one). Why?

Exercise 7.7

KEY CONCEPTS - samples, bias

a) How many individuals responded to the question? Those individuals represent the sample.

b) What population is the sample from? Does the sample represent an SRS from this population? What are some possible sources of bias in this example? Are sources of bias reduced by taking larger samples?

Exercise 7.9

KEY CONCEPTS - selecting an SRS with a table of random numbers

The table of random numbers can be used to select an SRS of numbers. In order to use it to select a random sample of six minority managers, the managers need to be assigned numerical labels. So that everyone does the problem the "same" way, we have first labeled the managers according to alphabetical order in the list.

01 - Abdulhamid	08 - Duncan	15 - Huang	22 - Puri
02 - Agarwal	09 - Fernandez	16 - Kim	23 - Richards
03 - Baxter	10 - Fleming	17 - Liao	24 - Rodriguez
04 - Bonds	11 - Gates	18 - Mourning	25 - Santiago
05 - Brown	12 - Goel	19 - Naber	26 - Shen
06 - Castillo	13 - Gomez	20 - Peters	27 - Vega
07 - Cross	14 - Hernandez	21 - Pliego	28 - Wang

If you go to line 139 in the table and start selecting two digit numbers, then you should get the same answer as given in the complete solution. If your entire sample has not been selected by the end of line 139, continue to the next line in the table.

The sample consists of the managers ______________________________

Exercise 7.12

KEY CONCEPTS - stratified random sample

What are the two strata from which you are going to sample? A stratified random sample consists of taking an SRS from each stratum and combining these to form the full sample. How large an SRS will be taken from each of the two strata? How would you label the units in each of the two strata from which you will sample? Fill in the following table to describe your sampling plan.

	STRATA	
	Midsize accounts	Small accounts
Number of units in stratum		
Sample size		
Labeling method		

In practice, you would probably use the table of random numbers to first select the SRS from the midsize accounts and then you would select the SRS from the small accounts. In this problem you are not going to select the full samples, but only the first five units from each stratum. Start in Table B at line 115 and first select 5 midsize accounts and then continue in the table to select 5 small accounts. Write the numerical labels below.

First five midsize accounts ______________________________.

First five small accounts ______________________________.

Exercise 7.22

KEY CONCEPTS - populations and sources of bias

What variable was measured and what was the sample? Now, try and identify the population as exactly as possible. Where the information is not complete, you may need to make assumptions to try to describe the population in a reasonable way. Make sure not to confuse the population of interest with the population actually sampled. When they don't coincide there is always a strong potential for bias. What are some possible sources of bias in this example?

Exercise 7.25

KEY CONCEPTS - populations, samples, bias

What is the population of interest to the Miami Police Department? Is there a problem with undercoverage in the way the addresses for the sample are selected? How might this bias the results? Could there be response bias in this study as well? How might this bias the results?

Exercise 7.29

KEY CONCEPTS - selecting an SRS with a table of random numbers

The table of random numbers can be used to select an SRS of blocks. In order to use it to sample blocks from the census tract, the blocks need to be assigned numerical labels. The numbering system for the 44 blocks is 1000, 1001, ..., 1005, 2000, 2001, ..., 2011, 3000, 3001, ..., 3025. You could represent these 44 blocks by their four-digit numbers and go to Table B, but because only 44 of the 10,000 four-digit numbers correspond to blocks, most four-digit numbers encountered in the table would not correspond to blocks. As an alternative, the blocks could be renumbered from 01 to 44 according to the following scheme. The two-digit numbers assigned to each block are in parentheses.

1000(01), 1001(02), ..., 1005(06)
2000(07), 2001(08), ..., 2011(18)
3000(19), 3001(20), ..., 3025(44)

Enter Table B at line 125 and select two digit numbers using digits from 01 to 44 using the labeling scheme above. If your entire sample has not been selected by the end of line 125, continue to the next line in the table.

The sample consists of the blocks numbered ______________________________

Exercise 7.36

KEY CONCEPTS - systematic sampling

a) This is like the example except there are now 200 addresses instead of 100, and the sample size is now 5 instead of 4. With these two changes, you need to think about how many different systematic samples there are. Two different systematic samples are:

systematic sample 1 = 01, 41, 81, 121, 161
systematic sample 2 = 02, 42, 82, 122, 162

....

How many systematic samples are there altogether? Choosing one of these systematic samples at random is equivalent to choosing the first address in the sample. The remaining four addresses follow automatically by adding 40. Carry this out using line 120 in the table.

b) Why are all addresses equally likely to be selected? First, how many systematic samples contain each address? The chance of selecting an address is the same as the chance of selecting the systematic sample that contains it. With this in mind, what is the chance of any address being chosen? By the definition of an SRS, all samples of 5 addresses are equally likely to be selected. In a systematic sample, are all samples of 5 addresses even possible?

Exercise 7.37

KEY CONCEPTS - sampling frame, undercoverage

a) Which households wouldn't be in the sampling frame? Make some educated guesses as to how these households might differ from those in the sampling frame (other than the fact that they don't have a phone number in the directory).

b) Random digit dialing makes the sampling frame larger — which households are added to it?

Exercise 7.41

KEY CONCEPTS - populations, samples, sample size, bias

a) What is the population of interest and the sample? Is the sample from the population of interest or are there potential sources of bias?

b) Is this a scientific sample using probability in the selection of the sample? Is the sample size large?

COMPLETE SOLUTIONS

Exercise 7.1

The explanatory variable is whether or not the person made use of handheld cellular phones (we assume on a regular basis) and the response is whether or not the person contracted brain cancer. No attempt was made to decide which individuals were going to make use of a cellular phone (the treatment), so this is an observational study, not an experiment.

Exercise 7.7

a) The sample are those that responded to the question and there are a total of 14,793 in this sample.

b) At best the population might be characterized as those individuals that go to this particular website. Even for this population, it is only a voluntary response sample as those with little interest in this question would not necessarily vote. Those with strong opinions tend to vote and are not necessarily representative of the population. Also males make heavier use of the web and may be overrepresented. Larger sample sizes cannot correct a poorly designed survey which contains many sources of bias.

Exercise 7.9

To choose a SRS of 6 managers to be interviewed, first label the members of the population by associating a 2 digit number with each.

01 - Abdulhamid	08 - Duncan	15 - Huang	22 - Puri
02 - Agarwal	09 - Fernandez	16 - Kim	23 - Richards
03 - Baxter	10 - Fleming	17 - Liao	24 - Rodriguez
04 - Bonds	11 - Gates	18 - Mourning	25 - Santiago
05 - Brown	12 - Goel	19 - Naber	26 - Shen
06 - Castillo	13 - Gomez	20 - Peters	27 - Vega
07 - Cross	14 - Hernandez	21 - Pliego	28 - Wang

Now enter Table B and read two-digit groups until 6 managers are chosen. Starting at line 139

55|58|8 9|94|04| 70|70|8 4|10|98 43|56|3 5|69|34| 48|39|4 5|17|19| 12|97|5 1|32|58| 13|04|8

The selected sample is 04 - Bonds, 10 - Fleming, 17 - Liao, 19 - Naber, 12 - Goel, and 13 - Gomez.

Exercise 7.12

There are 500 midsize accounts. We are going to sample 5% of these which is 25. You should label the accounts 001, 002, ..., 500 and select a SRS of 25 of the midsize accounts. There are 4400 small accounts. We are going to sample 1% of these which is 44. You should label the accounts 0001, 0002, ..., 4400 and select a SRS of 44 of the small accounts.

Starting at line 115, we first select 5 midsize accounts, that is an SRS of size 5 using the labels 001 through 500. Continuing in the table we select 5 small accounts, that is an SRS of size 5 using the labels 0001 through 4400. Note that for the midsize accounts we read from Table B using three digit numbers, and for the small accounts we read from the Table using four digit numbers.

610|41 7|768|4 94|322| 247|09 7|3698| 1452|6 318|93 32|59 1|4459| 2605|6 314|24
80|371 6|

The first five midsize accounts are those with labels 417, 494, 322, 247, and 097. Continuing in the table, using four digits instead of three, the first five small accounts are those with labels 3698, 1452, 2605, 2480 and 3716.

Exercise 7.22

The variable being measured is approval of the president's overall job performance which is recorded as approve or don't approve. The sample is the 1210 adults that were actually interviewed. The population of interest is probably all adult citizens of the U.S. or possibly just registered voters.

There are several possible sources of bias in the study. The states of Alaska and Hawaii were omitted and there is no reason to believe that the adult residents of these states were not intended to be part of the population (they may not have been included in the sample due to the higher cost of calling residents of these states). Any systematic differences in the opinions of the adults in Alaska and Hawaii and the remaining states will bias the results. Also, only residents with phones could be contacted and if the phone numbers were selected from phone books then residents with unlisted numbers could not be in the sample. This is another possible source of bias, which is just any systematic error in the way the sample represents the population. Finally, there may be bias due to nonresponse, as all adults contacted by phone may not have been willing to give their opinion.

Exercise 7.25

The population of interest is black residents of Miami. The sample is the black residents at the 300 mailing addresses chosen by the police. (If a non-black resident lives at one of these addresses, their opinions would not be included in the sample, as they are not part of the population).

The police are interviewing residents at addresses in predominantly black neighborhoods. The opinions of those black residents who do not live in predominantly black neighborhoods could not be included in the sample, resulting in undercoverage. It is likely that black residents in other neighborhoods may feel differently about the police, creating a potential source of bias if the population of interest is really *all* black residents. These black residents of Miami may be more satisfied with police service.

A more serious source of bias is response bias. It is unlikely that an adult will be honest when giving his opinion of police service to a police officer. They will probably indicate a greater degree of satisfaction than they really feel.

Exercise 7.29

The blocks have been renumbered from 01 to 44 according to the following scheme:

1000(01), 1001(02), ..., 1005(06)
2000(07), 2001(08), ..., 2011(18)
3000(19), 3001(20), ..., 3025(44)

Starting at line 125 in Table B and ignoring repeats

96746 12149 37823 71868 18442 35119 62103 39244

the five two-digit numbers selected are 21 (block 3002), 18 (block 2011), 23 (block 3004), 19 (block 3000), and 10 (block 2003).

Exercise 7.36

a) We want to select 5 addresses out of 200, so we think of the 200 addresses as forty lists, each containing 5 addresses. We choose one address from the first 40, and then every 40th address after that. The first step is to go to Table B, line 120 and choose the first two digit random number you encounter that is one of the numbers 01, ..., 40.

35476

The selected number is 35, so the sample includes addresses numbered 35, 75, 115, 155, and 195.

b) Each individual is in exactly one systematic sample, and the systematic samples are equally likely to be chosen. In our previous example, there were 40 systematic samples, each containing 5 addresses. The chance of selecting any address is the chance of picking the systematic sample that contains it, which is 1 in 40.

A simple random sample of size n would allow every set of n individuals an equal chance of being selected. Thus, in this exercise, when using a SRS the sample consisting of the addresses numbered 1, 2, 3, 4, and 5 would have the same probability of being selected as any other set of 5 addresses. For a systematically selected sample, all samples of size n do not have the same probability of being selected. In our exercise the sample consisting of the addresses numbered 1, 2, 3, 4, and 5 would have zero chance of being selected since the numbers of the addresses do not all differ by 40. The sample we selected in (a), 35, 75, 115, 155, and 195 had a 1 in 40 chance of being selected, so all samples of five addresses are not equally likely.

Exercise 7.37

a) Households omitted from the frame are those which do not have a telephone number listed in the telephone directory. The types of people who might be underrepresented are poorer people (including the homeless) who cannot afford to have a phone, and the group of people who have unlisted numbers. It is harder to characterize this second group. As a group they would tend to have more money as you need to pay to have your phone number unlisted or it might include more single women who do not want their phone numbers available and possibly people whose jobs put them in contact with large groups of people who might harass them if their phone number was easily accessible.

b) People with unlisted numbers will be included in the sampling frame. The sampling frame would now include any household with a phone. One interesting point is that all households will not have the same probability of getting in the sample, as some households have multiple phone lines and will be more likely to get in the sample. So, strictly speaking, random digit dialing will not actually provide a SRS of households with phones. Just a SRS of phone numbers!

Exercise 7.41

a) The population is all people who live in Ontario. Because everyone uses health care, it is not restricted to adults, etc. The sample is the 61,239 residents of Ontario who were interviewed.

b) Yes. This is a very large sample and it is a probability sample, so we expect that the sample proportions are quite close to the population proportions.

CHAPTER 8
PRODUCING DATA: EXPERIMENTS

OVERVIEW

Experiments are studies in which one or more **treatments** are imposed on experimental **units** or **subjects**. A treatment is a combination of **levels** of the explanatory variables, called **factors**. The design of an experiment is a specification of the treatments to be used and the manner in which units or subjects are assigned to these treatments. The basic features of well-designed experiments are **control**, **randomization**, and **replication**.

Control is used to avoid confounding (mixing up) the effects of treatments with other influences such as lurking variables. One such lurking variable is the **placebo effect**, which is the response of a subject to the fact of receiving any treatment. The simplest form of control is **randomized comparative experimentation** which involves comparisons between two or more treatments. One of these treatments may be a **placebo** (fake treatment), and those subjects receiving the placebo are referred to as a **control group**.

Randomization can be carried out using the ideas learned in Chapter 7 of your text. Randomization is carried out before applying the treatments and helps control bias by creating treatment groups that are similar. Replication, the use of many units in an experiment, is important because it reduces the chance variation between treatment groups arising from randomization. Using more units helps increases the ability of your experiment to establish differences between treatments.

Further control in an experiment can be achieved by forming experimental units into **blocks** that are similar in some way which is thought to affect the response, similar to strata in a stratified sample design. In a **block design**, units are first formed into blocks and then randomization is carried out separately in each block. **Matched pairs** are a simple form of blocking used to compare two treatments. In a matched pairs experiment either the same unit (the block) receives both treatments in a random order or very similar units are matched in pairs (the blocks). In the latter case, one member of the pair receives one of the treatments and the other member the remaining treatment. Members of a matched pair are assigned to treatments using randomization.

Some additional problems which can occur that are unique to experimental designs are lack of **blinding** and **lack of realism**. These problems should be addressed when designing the experiment.

GUIDED SOLUTIONS

Exercise 8.3

KEY CONCEPTS - identifying experimental units, factors, treatments, and response variables

a) You need to read the description of the study carefully. To identify the "individuals," ask yourself, what are the treatments going to be applied to? What is being measured on these individuals? This is the response.

b) Draw a diagram like that in Figure 8.1 of your text in the space below. What are the factors and how many levels do they have? How many combinations of the levels of the factors are there? This is the number of treatments.

c) How many individuals are being used for each treatment? Multiply this by the number of treatments to obtain the number of individuals required for the experiment.

Exercise 8.5

KEY CONCEPTS - completely randomized design, randomization

How many treatment groups are there and how many rats are assigned to each group? This information must be included in your outline. Now, using Figure 8.3 of your text as a model, give a diagram to outline a completely randomized design for the study, using appropriate labels in the diagram.

Now, label the rats 01 to 18. Enter Table B at line 142 and select an SRS of 6 rats to receive black tea extract. Continue in Table B, selecting 6 more to receive green tea extract. The remaining 6 receive the placebo.

Six rats assigned to receive black tea extract ______________________________

Six rats assigned to receive green tea extract ______________________________

Six rats assigned to receive placebo ______________________________

Exercise 8.7

KEY CONCEPTS - randomized comparative experiments, observational studies

Is the new design suggested by the executive an experiment? What are the disadvantages of their approach?

Exercise 8.11

KEY CONCEPTS - double blind experiments, bias

What does it mean for the ratings to be blind? Was this done in this case? If not, how can this bias the results and in which direction?

Exercise 8.12

KEY CONCEPTS - matched pairs design, randomization

The first thing you should do is identify the treatments and the response variable. Next, decide what are the matched pairs in this experiment. How will you use a coin flip to assign members of a pair to the treatments? What will you measure and how will you decide whether the right-hand tends to be stronger in right-handed people?

Exercise 8.23

KEY CONCEPTS - completely randomized design, factor levels, response variable

a) How many subjects are in the study? Identify them as well as the response measured on each subject. What is the factor being studied and what are its levels?

b) Using Figure 8.3 of your text as a model, give a diagram to outline a completely randomized design for the study, using appropriate labels.

Exercise 8.27

KEY CONCEPTS - completely randomized design, practical difficulties

a) Using Figure 8.3 of your text as a model, give a diagram to outline a completely randomized design for the study, using appropriate labels.

b) How long would you need to carry out the experiment? How would you assign subjects to the different treatment groups? Do you anticipate any practical or ethical difficulties in doing this? Explain.

Exercise 8.29

KEY CONCEPTS - completely randomized design, randomization

The list of names of the 36 students has been reproduced below. Assign a numerical label to each. Be sure to use the same number of digits for each label. In the complete solutions, the subjects are labeled in alphabetical order starting at 00 and ending at 35. If you want your answer to agree with the complete solutions, then you will need to use the same labeling scheme.

Alomar	Denman	Han	Liang	Padilla	Valasco
Asihiro	Durr	Howard	Maldonado	Plochman	Vaughn
Bennett	Edwards	Hruska	Marsden	Rosen	Wei
Bikalis	Farouk	Imrani	Montoya	Soloman	Wilder
Chao	Fleming	James	O'Brian	Trujillo	Willis
Clemente	George	Kaplan	Ogle	Tullock	Zhang

Now start reading line 130 in Table B. Read across the row in groups of digits equal to the number of digits you used for your labels (for example, if you used two digits for labels, read line 130 in pairs of digits). You will need to keep reading until you have selected all the names for the first treatment. This may require you to continue on to line 131, line 132, and subsequent lines. After you have selected the names for treatment 1, continue in Table B to assign the six people to receive each of treatment 2, 3, 4, and 5. The remaining names are assigned to treatment 6.

Treatment 1 ________________________________

Treatment 2 ________________________________

Treatment 3 ________________________________

Treatment 4 ________________________________

Treatment 5 ________________________________

Treatment 6 ________________________________

Exercise 8.37

KEY CONCEPTS - blind experiments, matched pairs experiments

Should subjects be told which two hamburger chains are being compared? What are the matched pairs in this experiment? How should the randomization be carried out?

COMPLETE SOLUTIONS

Exercise 8.3

a) The individuals are the different batches, and the response variable is the yield of the chemical reaction.

b) There are two factors in the experiment — temperature and stirring rate. The treatments are the different temperature and stirring rate combinations. Because there are two levels of temperature and three levels of stirring rate, there are a total of six treatments. The diagram below lays out the different treatment combinations in the design.

Factor A Temperature	Factor B Stirring Rate: 60 rpm	90 rpm	120 rpm
50 deg	1	2	3
60 deg	4	5	6

c) There are six treatments with two batches of the product at each treatment, so 12 batches or "individuals" are needed for the experiment.

Exercise 8.5

There are three treatments, black tea extract, green tea extract and a placebo. There are 18 rats for the study, so six rats are assigned at random to each of the treatment groups.

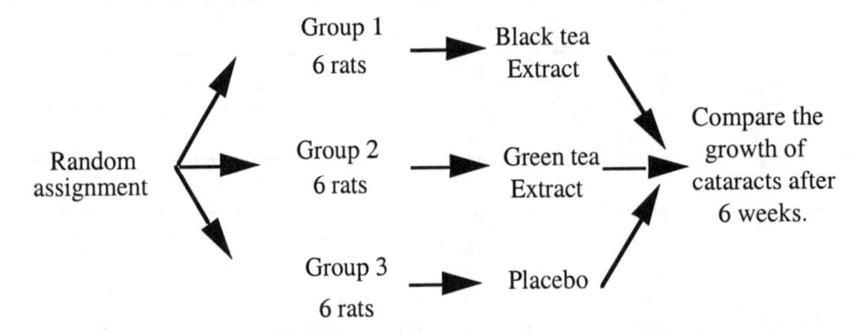

We have reproduced Table B starting at line 142 below. Read across the row in groups of two digits since you used two digits for your labels, skipping repeats. The first six underlined numbers correspond to the rats receiving black tea extract and the next six to those receiving green tea extract. The remaining rats receive the placebo.

```
72829 50232 97892 63408 77919 44575 24870 04178
88565 42628 17797 49376 61762 16953 88604 12724
62964 88145 83083 69453 46109 59505 69680 00900
19687 12633 57857 95806 09931 02150 43163 58636
37609 59057 66967 83401 60705 02384 90597 93600
54973 86278 88737 74351 47500 84552 19909 67181
00694 05977 19664 65441 20903 62371 22725 53340
71546 05233 53946 68743 72460 27601 45403 88692
07511 88915 41267
```

Six rats assigned to receive black tea extract	02, 08, 17, 10, 05 and 09.
Six rats assigned to receive green tea extract	06, 16, 01, 07, 18 and 15.
Six rats assigned to receive placebo	03, 04, 11, 12, 13, and 14.

Exercise 8.7

The new design is an observational study, not an experiment. The electric company wants the only systematic differences in groups to be the treatments. Electric use varies from year to year depending on the weather. If charts or indicators are introduced in the second year, and the electric consumption in the first year is compared with the second year, you won't know if the observed differences are due to the introduction of the chart or lurking variables. For example, if the comparison is being made in the summer months, it is possible that the second year had a cooler summer, which reduced the need for air conditioning and reduced electric consumption, rather than the introduction of charts or indicators. A control group ensures that influences other than the introduction of the indicators or charts operate equally on all groups.

Exercise 8.11

The ratings were not blind because the experimenter who rated the subject's level of anxiety presumably knew whether they were in the meditation group or not. Since the experimenter was hoping to show that those that meditated had lower levels of anxiety, he might unintentionally rate those in the meditation group as having lower levels of anxiety, if there is any subjectivity in the ratings. It would be better if a third party who did not know which group the subjects belonged to rated the subjects anxiety levels.

Exercise 8.12

We have 10 subjects available. There are two treatments in the study. Treatment 1 is squeezing with the right hand and treatment 2 is squeezing with the left hand. The response is the force exerted as indicated by the reading on the scale.

To do the experiment, we use a matched pairs design. The matched pairs are the two hands of a particular subject. We should randomly decide which hand to use first, perhaps by flipping a coin. We measure the response for each hand and then compare the forces for the left and right hands over all subjects to see if there is a systematic difference between the two hands.

Exercise 8.23

a) The experimental subjects are the 22,000 male physicians participating in the study. The factor is the drug taken at two levels—aspirin and placebo. The response measured was whether or not the subject had a heart attack during the several years that the study was conducted.

b) The diagram below outlines the experiment.

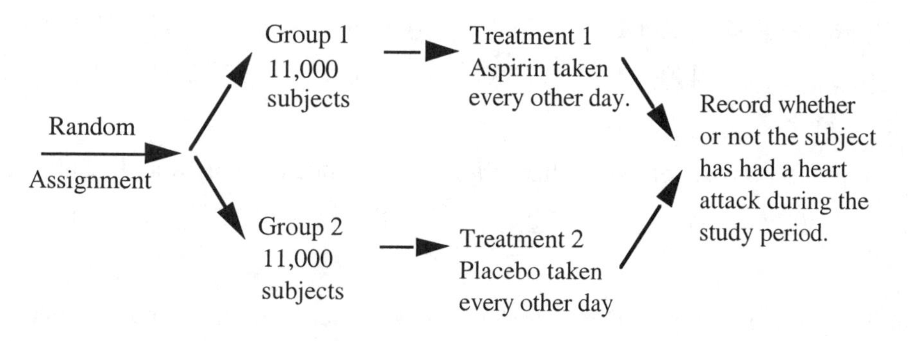

Exercise 8.27

a) The diagram below outlines the experiment.

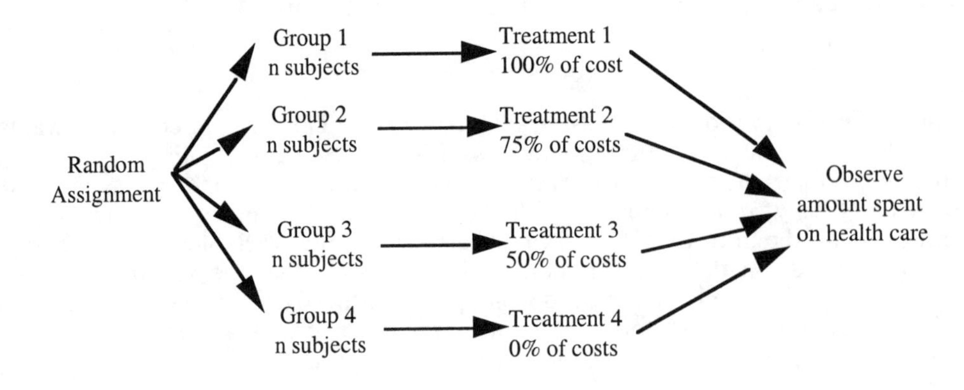

b) As a practical issue, it may take a long time to carry out the study. The experimenters will need to wait until enough claims have been filed to obtain the information they need. Ethically, we are going to assign subjects to different insurance plans; some people might object to this because some will be required to pay for part of their health care while others will not.

Exercise 8.29

00 Alomar
01 Asihiro
02 Bennett
03 Bikalis
04 Chao
05 Clemente
06 Denman
07 Durr
08 Edwards
09 Farouk
10 Fleming
11 George
12 Han
13 Howard
14 Hruska
15 Imrani
16 James
17 Kaplan
18 Liang
19 Maldonado
20 Marsden
21 Montoya
22 O'Brian
23 Ogle
24 Padilla
25 Plochman
26 Rosen
27 Soloman
28 Trujillo
29 Tullock
30 Valasco
31 Vaughn
32 Weo
33 Wilder
34 Willis
35 Zhang

Line 130 from Table B is reproduced below. We should read line 130 in pairs of digits from left to right. We have placed vertical bars between consecutive pairs to indicate how we have read the table. We underline those pairs that correspond to labels in our list and that have not been previously selected. On line 130 we have

69|05|1 6|48|17| 87|17|4 0|95|17| 84|53|4 0|64|89| 87|20|1 9|72|45

We only find 5 of our labels so we need to continue reading on line 131.

05|00|

We now have six labels. The treatment 1 group consists of Clemente, James, Kaplan, Marsden, Maldonado, and Alomar. Continuing in line 131 (and ignoring pairs corresponding to previously selected subjects) we assign the next six subjects to treatment 2.

7 1|66|32| 81|19|4 1|48|73| 04|19|7 8|55|76| 45|19|5 9|65|65

68|73|2 5|52|59| 84|29|2 0|87|96| 43|16|5 9|37|39| 31|68|5 9|71|50

45|74|0 4|18|

The treatment 2 group consists of Wei, Chao, Plochman, Tullock, Vaughn and Liang. Continuing,

07| 65|56|1 3|33|02| 07|05|1 9|36|23| 18|13|2 0|95|47

27|

the treatment 3 group consisting of Durr, Howard, Wilder, Bennett, Ogle, and Soloman. Continuing

81|6 7|84|16| 18|32|9 2|13|37| 35|21|3 3|77|41| 04|31|2 6|85|08

66|92|5 5|56|58| 39|10|0 7|84|58| 11|

the treatment 4 group consists of Zhang, Montoya, Rosen, Edwards, Fleming, and George. Again, continuing,

```
                                                20|6 1|98|76| 87|15|1 3|12|60
08|42|1 4|47|53| 77|37|7 2|87|44| 75|59|2 0|85|63| 79|14|0 9|24|54
53|64|5 6|68|12| 61|42|1 4|78|36| 12|60|9 1|53|73| 98|48|1 1|45|92
66|83|1 6|89|08| 40|77|2 2|
```

the treatment 5 group consists of Imrani, Han, Hruska, Farouk, Padilla and O'Brian. The six remaining subjects, Ashiro, Bikalis, Denman, Trujillo, Valasco and Willis are assigned to treatment 6. Using Table B can become tedious with a large number of subjects and it is best to leave such calculations to a computer.

Exercise 8.37

Subjects should not be told which burger comes from Wendy's or McDonald's and in fact shouldn't be told which two burger chains are being compared as identification of the burger might be easy. It would be best to make the hamburgers in such a way that they look alike in terms of size, bun, and condiments. Each subject would be presented with the two burgers in a random order. The randomization of the order should be done by a flip of a coin for each subject.

CHAPTER 9

INTRODUCING PROBABILITY

OVERVIEW

Probability is a branch of mathematics that studies random processes over a long series of repetitions. A process or phenomenon is called **random** if its outcome is uncertain. Although individual outcomes are uncertain, when the process is repeated a large number of times the underlying distribution for the possible outcomes begins to emerge. For any outcome, its **probability** is the proportion of times, or the relative frequency, with which the outcome would occur in a long series of repetitions of the process. It is important that these repetitions or trials be **independent** for this property to hold. By independence we mean that the outcome of one trial must not influence the outcome of any other.

You can study random behavior by carrying out physical experiments such as coin tossing or die rolling, or you can simulate a random phenomenon on the computer. Using the computer is particularly helpful when we want to consider a large number of trials. Randomness and independence are the keys to using the rules of probability.

The description of a random phenomenon begins with the **sample space, S**, which is the list of all possible outcomes. A set of outcomes is called an **event.** Once we have determined the sample space, a **probability model** tells us how to assign probabilities to the various events that can occur. There are four basic rules that probabilities must satisfy.

- Any probability is a number between 0 and 1. $P(A)$ means "the probability of A." If the probability is 0, the event will never occur. If the probability is 1, the event will always occur.

- All possible outcomes together must have probability 1.

- The probability that an event does not occur is 1 minus the probability that the event occurs. Using our notation, this can be written as

$$P(A) = 1 - P(\text{not } A)$$

- If two events have no outcomes in common, the probability that one or the other occurs is the sum of their individual probabilities. These events are **disjoint.** This is the addition rule for disjoint events, namely

$$P(A \text{ or } B) = P(A) + P(B)$$

In a sample space with a finite number of outcomes, probabilities are assigned to the individual outcomes and the probability of any event is the sum of the probabilities of the outcomes it contains. All outcomes must have a probability between 0 and 1; the sum of all probabilities must add up to 1.

Even a sample space with an infinite number of outcomes can have probabilities assigned to it. To do this we use a density curve. (Refer back to Chapter 3 of your textbook to refresh your memory.) The area under a density curve must be equal to 1. Probabilities are assigned to events as areas under the density curve. The normal distribution is the most useful of the density curves. **Normal distributions are probability models.**

A **random variable** is a variable whose value is a numerical outcome of a random phenomenon. The **probability distribution** of a random variable tells us about the possible values of the random variable and how to assign probabilities to these values. A random variable can be **discrete** or **continuous**. A **discrete random variable** has finitely many possible values. Its distribution gives the probability of each value. A **continuous random variable** takes all values in some interval of numbers. A **density curve** describes the distribution of a continuous random variable.

GUIDED SOLUTIONS

Exercise 9.3

KEY CONCEPTS - estimating probabilities as the proportion of times an event occurs in many repeated trials

Record the number of heads for your 50 spins.

The probability of an event is the proportion of times the event occurs in many repeated trials of a random phenomenon. To estimate the probability of a head when you spin a nickel, determine the proportion of times this occurred in your 50 trials.

Estimate =

Exercise 9.10

KEY CONCEPTS - sample space

One of the main difficulties encountered when describing the sample space is finding some notation to express ideas formally. Following the text, our general format is $S = \{ \quad \}$, where a description of the outcomes in the sample space is included within the braces.

a) We know that the weights of healthy adult women are positive numbers, but it is not clear what the smallest and largest possible weights should be. If you say that 90 pounds is the smallest possible weight a healthy adult woman can have, why not 89.99 pounds? If you say that the highest possible weight a healthy adult woman can have is 300 pounds, why not 300.01 pounds? The point is that the best we probably can do is say that healthy adult women weigh more than 0 pounds and take the sample space to be

S = {set of all numbers greater than zero}

b) S =

c) S =

Exercise 9.29

KEY CONCEPTS - probabilities in a finite sample space, legitimate probability models

a) In a legitimate probability model, the probabilities of all possible outcomes in the sample space must add to 1. The six colors listed and the outcome "the vehicle you choose has a color other than the six listed" account for all possible outcomes. Add up the six probabilities given in the table. What must the probability be that the vehicle you choose has a color other than the six listed?

Sum of the six probabilities listed in the table =

Probability the vehicle you choose has a color other than the six listed =

b) These events are disjoint. What do our rules of probability tell us about the probability that one or the other occurs?

Exercise 9.33

KEY CONCEPTS - sample spaces for simple random sampling, probabilities of events

a) It's easy to make the list. S = {(Abby, Deborah), (Abby, Mei-Ling), etc.}. There is no need to include both (Abby, Deborah) and (Deborah, Abby) in your list, since both refer to the same two individuals.

b) How many outcomes are there in the sample space in part (a)? If they are equally likely, what is the probability of each?

c) How many outcomes in S include Mei-Ling? When the outcomes are equally likely, the probability of the event is just

$$\frac{\text{number of outcomes in the event}}{\text{number of outcomes in } S}$$

d) How many outcomes in S include neither of the two men?

Exercise 9.40

KEY CONCEPTS - random numbers, density curve

a) Recall that a discrete random variable has finitely many possible values. A continuous random variable takes all values in some interval of numbers. Which is true in this example?

b) The total area underneath any density curve is 1. Since the density curve has constant height over the range 0 to 2, what must the height be to make the area under this curve equal to 1? Note that since the area under the density curve is a rectangular region, the formula for the area of a rectangle will be useful.

Now draw a graph of the density curve.

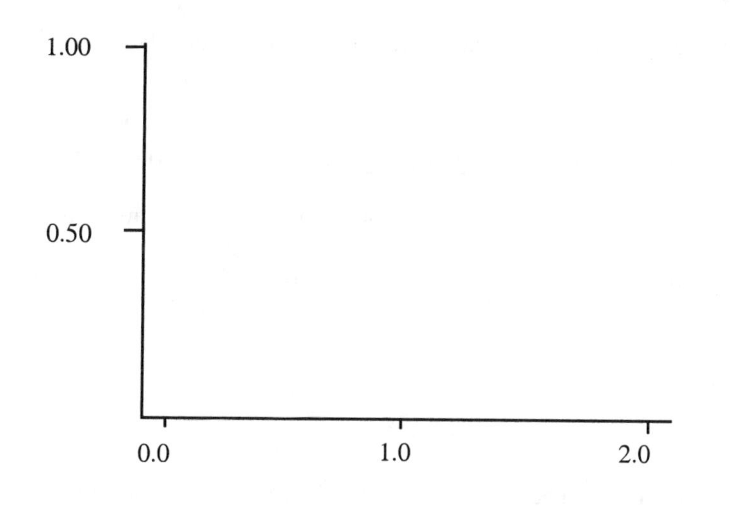

c) $P(Y \leq 1)$ is the area under the density curve below 1. You may find it helpful to sketch this region on your graph in part (b) This is a rectangular region, so use the formula for the area of a rectangle to compute this area.

Exercise 9.43

KEY CONCEPTS - normal random variables

a) Y stands for the score of a randomly chosen student. The event of interest is "the student has a score above 300." Because Y represents the student's score, in terms of Y the event is

$$\{Y > 300\}$$

To compute the probability of this event, recall that the mean is 300. What is the area under a normal curve corresponding to the region greater than the mean?

b) Write the event in terms of Y. To compute the probability, first determine how many multiples of the standard deviation 35, 370 is above the mean? Now apply the 68-95-99.7 rule.

Exercise 9.50

KEY CONCEPTS - simulating a random phenomenon

a) You will need to use your computer software to simulate the 100 trials. As the problem suggests, the key phrase to look for in your software is "Bernoulli trials." Your software may allow you to actually generate the words "Hit" and "Miss" or perhaps the letters H and M. More likely, your software will allow you to generate only the numbers 0 and 1. In this case, count a 0 as a miss and a 1 as a hit.

After you do this, the computer can be used to calculate the proportion of hits.

Proportion of hits =

For most students, the proportion of hits will be within 0.05 or 0.10 of the true probability of 0.5.

b) You need to go through your sequence to determine the longest string of hits or misses.

Longest run of shots hit =   Longest run of shots missed =

COMPLETE SOLUTIONS

Exercise 9.3

We spun a nickel 50 times and got 22 heads. From this, we estimate the probability of heads to be

Estimate = 22/50 = 0.44.

Your results will probably differ somewhat, but your estimate of the probability of heads will be the number of heads you got in 50 spins divided by 50.

Exercise 9.10

a) S = {set of all numbers greater than zero}.
b) S = {0, 1, 2, ..., 11,000}.
c) S = {0, 1, ..., 12}.

Exercise 9.29

a) Sum of the six probabilities listed in the table = 0.765.

The sum of all possible outcomes must be 1, so the probability that the vehicle you choose has a color other than the six listed, when added to 0.765, must make the sum 1. Thus,

Probability the vehicle you choose has a color other than the six listed = 1 – 0.765 = 0.235

b) Because these events are disjoint, we simply add the probabilities of silver and white to get

Probability that a randomly chosen vehicle is either silver or white = 0.210 + 0.156 = 0.366

Exercise 9.33

a) S = {(Abby, Deborah), (Abby, Mei-Ling), (Abby, Sam), (Abby, Roberto), (Deborah, Mei-Ling), (Deborah, Sam), (Deborah, Roberto), (Mei-Ling, Sam), (Mei-Ling, Roberto), (Sam, Roberto)}.

b) There are 10 possible outcomes. Since they are equally likely, each has probability 0.10.

c) Mei-Ling is in four of the outcomes, so her chance of attending the conference in Paris is 4/10 = 0.4.

d) The chosen group must contain two women, (Abby, Deborah), (Abby, Mei-Ling), or (Deborah, Mei-Ling). There are three possibilities, so the desired probability is 3/10 = 0.3.

Exercise 9.40

a) Here Y takes on all values in the interval 0 to 2. Thus, Y is continuous.

b) The density curve will be a rectangle with base covering the region 0 to 2. The base has length 2. The area of a rectangle is base × height, so the area of the density curve is 2 × height. This area must equal 1, hence the height must be 0.5.

Here is a graph of the density curve.

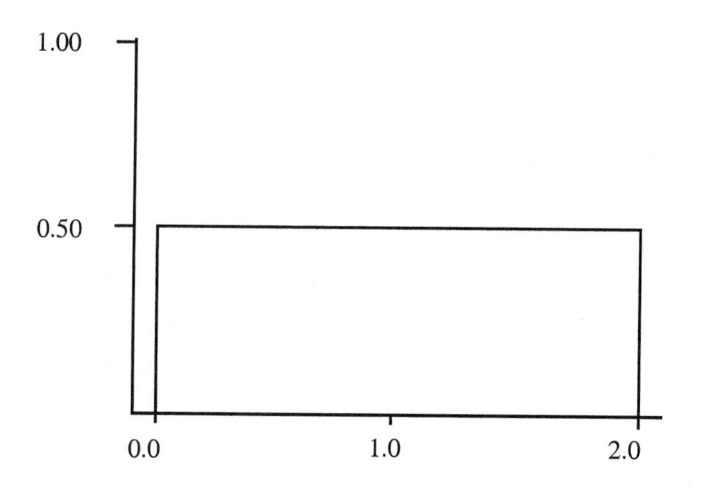

c) Here is a graph of the desired region.

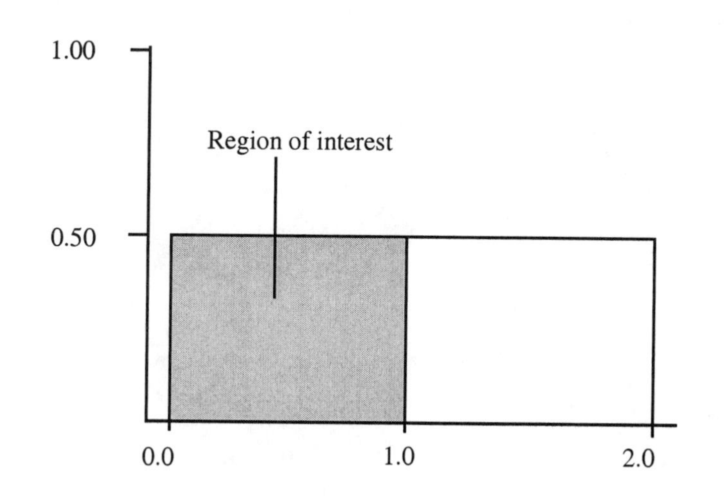

The region of interest is a rectangle with base of length 1.0 and height 0.5. Thus the area of this rectangle is $1 \times 0.5 = 0.5$, so

$$P(Y \leq 1) = 0.5$$

Exercise 9.43

a) In terms of Y, the event is $\{Y > 300\}$. Recall that half the area under the normal curve corresponds to the region greater than the mean and half to the region less than the mean. The mean in this case is 300, so

$$P(Y > 300) = \text{area under a normal curve greater than the mean} = 0.5$$

b) In terms of Y, the event is $\{Y > 370\}$. Notice that 370 is 70 above 300. Because 70 is equal to twice the standard deviation of 35, 370 is two standard deviations above the mean. The 68-95-99.7 rule tells us that 95% of the area under a normal curve is within two standard deviations of the mean. Because of the symmetry of the normal curve, this implies that 2.5% of the area is more than two standard deviations below the mean and 2.5% of the area is more than two standard deviations above the mean. Thus,

$$P(Y > 370) = \text{probability of being more than two standard deviations above the mean} = 0.025.$$

Exercise 9.50

a) Here is our sequence of hits (H) and misses (M).

H	H	M	H	H	H	M	M	H	H	H	M	H	M	H
M	H	M	M	H	M	M	H	H	H	M	M	H	H	H
M	M	M	M	H	M	M	H	H	H	H	M	H	H	H
M	M	M	M	M	H	M	H	H	M	H	M	H	M	M
H	H	H	H	M	H	M	M	M	M	H	H	M	H	H
M	M	H	H	H	M	M	H	H	M	M	H	M	H	M
M	H	M	H	H	M	H	H	H	H					

proportion of hits = .54

b) You need to go through your sequence to determine the longest string of hits or misses. In our example,

Longest run of shots hit = 4 (occurred more than once)

Longest run of shots missed = 5

CHAPTER 10

SAMPLING DISTRIBUTIONS

OVERVIEW

Statistical inference is the technique that allows us to use the information in a sample to draw conclusions about the population. Associated with any sample statistic is its **sampling distribution,** which is the distribution of values taken by the statistic in all possible samples of the same size from the same population. The sampling distribution can be described in the same way as the distributions you encountered in Chapters 1 and 2 of your textbook. Three important features are

- measure of center
- measure of spread
- description of the shape of the distribution

The statistic discussed first is the **sample mean**, $\bar{x}$, which is an estimate of the **population mean** μ. $\bar{x}$ has a number of convenient properties when taken from an SRS that allow one to make a variety of inferences about μ. As with all sampling distributions, we need to know the mean, standard deviation, and the shape of the distribution. In fact, we know from the **central limit theorem** that the shape of the distribution of the sample mean is very close to normal when simple random sampling is used and sample sizes are large. The **law of large numbers** also tells us more about the behavior of $\bar{x}$ as the sample size increases. The law of large numbers describes how the mean of many observations of a random process will get closer and closer to the mean of the population.

Here are the basic facts about the sample mean from an SRS of size n taken from a population where the mean is μ and the standard deviation is σ:

- $\bar{x}$ is an **unbiased estimate** of the population mean μ so the mean of $\bar{x}$ is the population mean μ
- The standard deviation is $\sigma/\sqrt{n}$, where σ is the population standard deviation. (This shows that there is less variation in averages than in individuals and that averages based on larger samples are less variable than those based on smaller samples.)
- The shape of the distribution of the sample mean depends on the shape of the population distribution. If the population was normal, $N(\mu, \sigma)$, then the sample mean has normal distribution $N(\mu, \sigma/\sqrt{n})$. Also, for a large sample the central limit theorem tells us that the sample mean has *approximately* a normal distribution $N(\mu, \sigma/\sqrt{n})$.

A statistic is called **unbiased** if the sampling distribution of the statistic is centered (has its mean) at the value of the population parameter. This means that the statistic tends to neither overestimate nor underestimate the parameter.

Statistical process control consists of methods for monitoring a process over time so that any changes in the process can be detected and corrected quickly. This is an economical method for maintaining product quality in a manufacturing process. We say that a process that continues over time is in **control** if it is operating under stable conditions. More precisely, the process is in control with respect to some variable measured on the process if the distribution of this variable remains constant over time.

Control charts are used to monitor the values of a variable measured on a process over time. One of the most common control charts is the $\bar{x}$ **control chart**. This is produced by observing a sample of n values of the variable of interest periodically and plotting the means $\bar{x}$ of these values versus the time order of the samples on a graph. A solid **centerline** at the target value of the process mean μ for the variable is drawn on the graph, as are dashed **control limits** at

$$\mu \pm 3\frac{\sigma}{\sqrt{n}}$$

where σ is the process standard deviation of the variable. This chart helps us decide if the process is in control with mean μ and standard deviation σ. The probability that the next point (value of $\bar{x}$) on such a chart lies outside the control limits is about 0.003 if the process is in control. Such a point would be evidence that the process is **out of control**, that is, that the distribution of the process has changed for some reason. When a process is deemed out of control, a cause for the change in the process should be sought.

GUIDED SOLUTIONS

Exercise 10.3

KEY CONCEPTS - statistics and parameters

In deciding whether a number represents a parameter or a statistic, you need to think about whether it is a number that describes a population of interest or whether it is a number computed from the particular sample that was selected. What is the population and what is the sample in the problem? Based on this, indicate whether the number is a parameter or a statistic.

2.5003 cm:

2.5009 cm:

Exercise 10.7

KEY CONCEPTS - standard deviation of the sampling distribution of $\bar{x}$

a) The key formula is that if $\bar{x}$ is the mean of an SRS of size n drawn from a large population with mean μ and standard deviation σ, then the standard deviation of the sampling distribution of $\bar{x}$ is $\frac{\sigma}{\sqrt{n}}$. To apply this here, identify σ and n and then complete the following:

Standard deviation of Juan's mean result = $\frac{\sigma}{\sqrt{n}}$ =

b) What value would n have to be so that $\frac{\sigma}{\sqrt{n}}$ is 5?

Write out your explanation of the advantage of reporting the average of several measurements rather than the result of a single measurement. Remember, don't use technical language.

Exercise 10.12

KEY CONCEPTS - central limit theorem

A sample size of 200 would be considered large enough to apply the central limit theorem. What does the central limit theorem say about the sampling distribution of $\bar{x}$? Fill in the blanks.

Sampling distribution of $\bar{x}$ is approximately $N(\quad , \quad)$

Now use the normal probability calculations you learned in Chapter 3 of your textbook to compute the probability that the mean $\bar{x}$ is greater than 2. If you have forgotten how to do these calculations, review the material in Chapter 3 of your textbook.

Exercise 10.19

KEY CONCEPTS - law of large numbers

Review the statement of the law of large numbers in the text. The mean payoff on a $1.00 bet is $0.947. What are the mean winnings for the gambler? This is μ in this problem. $\bar{x}$ would be the amount a gambler makes per bet on average after many bets on red. What does the law of large numbers say about $\bar{x}$?

Exercise 10.31

KEY CONCEPTS - sampling distributions, the central limit theorem

a) You may wish to refer to Chapter 1 of your textbook to refresh your memory about how to make a histogram. You may use the axes on the next page to assist you.

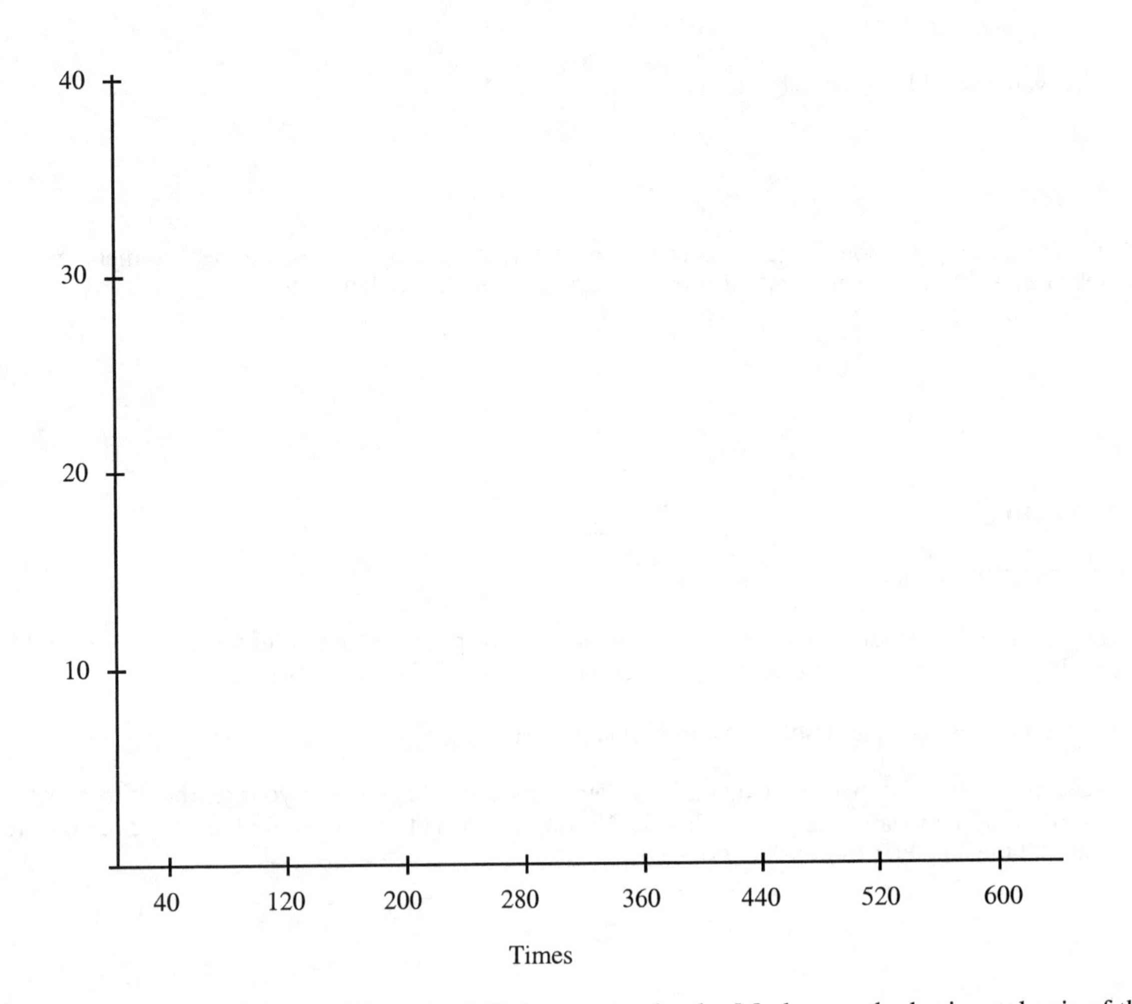

b) You might find it helpful to read Exercise 2.23 in your textbook. Mark μ on the horizontal axis of the histogram in part (a).

c) The 72 survival times are reproduced here to make you. Label these values from 01 to 72 and proceed to choose an SRS of size 12. Refer to Chapter 3 of your textbook if you need to review how to choose an SRS.

43	45	53	56	56	57	58	66	67
73	74	79	80	80	81	81	81	82
83	83	84	88	89	91	91	92	92
97	99	99	100	100	101	102	102	102
103	104	107	108	109	113	114	118	121
123	126	128	137	138	139	144	145	147
156	162	174	178	179	184	191	198	211
214	243	249	329	380	403	511	522	598

Line in Table B used to select SRS =

SRS of size 12 =

Sample mean $\bar{x}$ =

Remember to mark $\bar{x}$ on your histogram in part (a).

d) Write the results of your next four SRSs.

SRS 2 =

$\bar{x}$ =

SRS 3 =

$\bar{x}$ =

SRS 4 =

$\bar{x}$ =

SRS 5 =

$\bar{x}$ =

Remember to mark these four new values of $\bar{x}$ on your histogram in part (a).

Would you be surprised if all five $\bar{x}$'s fell on the same side of μ? Why?

e) Where would you expect the center of this sampling distribution to lie?

Exercise 10.33

KEY CONCEPTS - $\bar{x}$ control chart

You need to identify

$\mu =$
$\sigma =$
n = sample size =

From this information, determine

Center line = μ =

Control limits = $\mu \pm 3\dfrac{\sigma}{\sqrt{n}}$ =

Exercise 10.35

KEY CONCEPTS - natural tolerances

You need to identify

$\mu =$
$\sigma =$

From this information, determine

Natural tolerance = $\mu \pm 3\sigma =$

COMPLETE SOLUTIONS

Exercise 10.3

We desire information about the entire carload lot of ball bearings. The inspector chooses 100 bearings from the lot in order to decide whether to accept or reject the entire lot. In this problem, the entire carload lot is the population and the 100 bearings the sample. Thus

2.5003 cm. = a parameter since it describes the entire carload lot of bearings (the population).

2.5009 cm. = a statistic since it describes the sample of 100 bearings.

Exercise 10.7

a) In this problem $\sigma = 10$ and $n = 3$. Thus

$$\text{Standard deviation of Juan's mean result} = \frac{\sigma}{\sqrt{n}} = \frac{10}{\sqrt{3}} = 5.77$$

b) We want $5 = \dfrac{\sigma}{\sqrt{n}} = \dfrac{10}{\sqrt{n}}$. Solving this equation for $\sqrt{n}$, we must have

$$\sqrt{n} = 10/5 = 2$$

Squaring both sides yields $n = 4$.

Averages of several measurements are more likely to be closer to the true value of the quantity being measured than a single measurement. The magnitude of chance deviations or errors are smaller for averages than for individual observation.

Exercise 10.12

The sample size of 200 is reasonably large, so by the central limit theorem we might expect the sampling distribution of $\bar{x}$ to be approximately $N(1.6, 1.2/\sqrt{200}) = N(1.6, 0.085)$. Thus,

$$P(\bar{x} > 2) = P(\frac{\bar{x}-1.6}{0.085} > \frac{2-1.6}{0.085}) = P(z > 4.76)$$

which is approximately 0. (4.76 is well outside the range of values in our normal tables and so we know that this probability is much less than 0.0002; the probability that a standard normal random variable is greater than the largest value in the table, namely 3.49.)

Exercise 10.19

The gambler pays \$1.00 for an expected payout of \$0.947. His mean winnings are therefore \$0.947 – \$1.00 = –\$0.053 per bet. In other words, the expected losses of the gambler are \$0.053 per bet. The law of large numbers tells us that if the gambler makes a large number of bets on red, keeps track of his net winnings, and computes the average $\bar{x}$ of these, this average will be close to –\$0.053. In other words, he will find that he loses about 5.3 cents per bet on average.

Exercise 10.31

a) Here is a histogram of the 72 survival times.

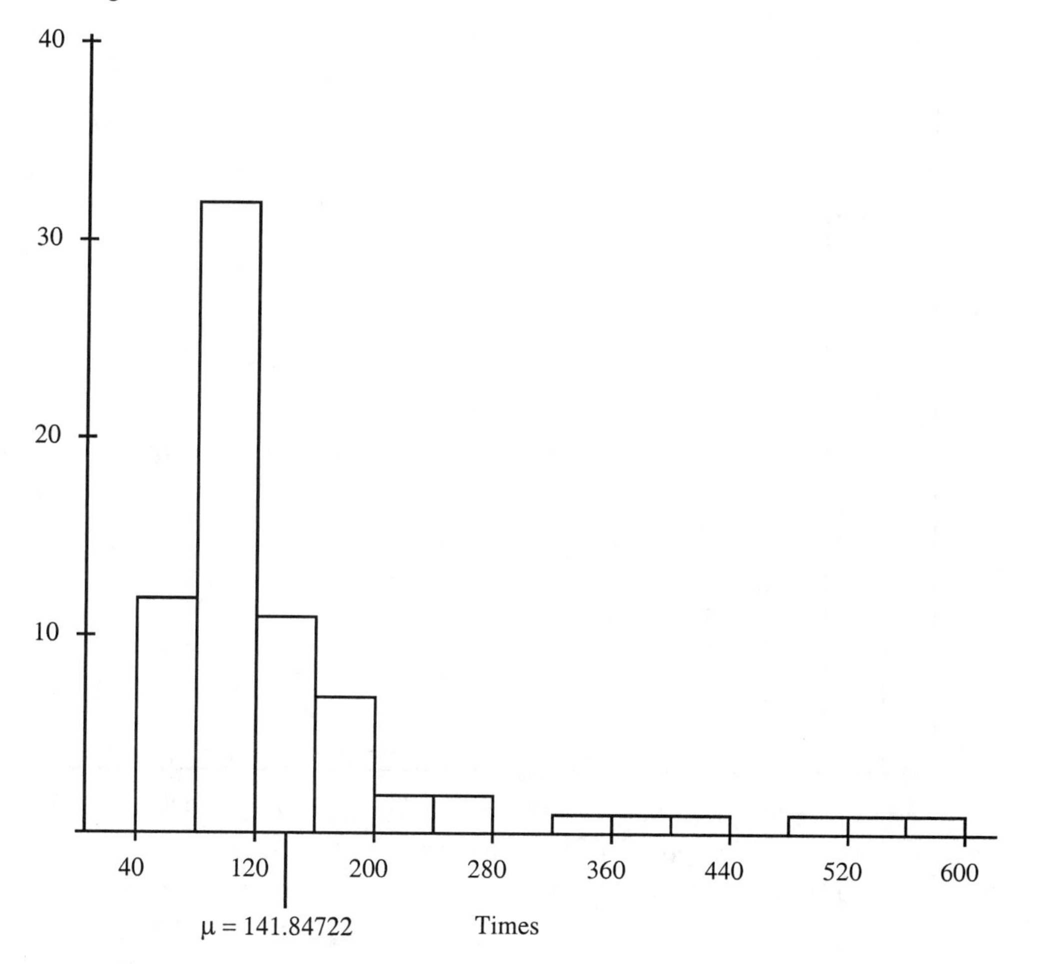

b) The mean is 141.84722. See the histogram in part (a) which has μ marked on the horizontal axis.

c) We label times from left to right. Labels are in parentheses.

(01) 43	(02) 45	(03) 53	(04) 56	(05) 56	(06) 57	(07) 58	(08) 66	(09) 67
(10) 73	(11) 74	(12) 79	(13) 80	(14) 80	(15) 81	(16) 81	(17) 81	(18) 82
(19) 83	(20) 83	(21) 84	(22) 88	(23) 89	(24) 91	(25) 91	(26) 92	(27) 92
(28) 97	(29) 99	(30) 99	(31) 100	(32) 100	(33) 101	(34) 102	(35) 102	(36) 102
(37) 103	(38) 104	(39) 107	(40) 108	(41) 109	(42) 113	(43) 114	(44) 118	(45) 121
(46) 123	(47) 126	(48) 128	(49) 137	(50) 138	(51) 139	(52) 144	(53) 145	(54) 147
(55) 156	(56) 162	(57) 174	(58) 178	(59) 179	(60) 184	(61) 191	(62) 198	(63) 211
(64) 214	(65) 243	(66) 249	(67) 329	(68) 380	(69) 403	(70) 511	(71) 522	(72) 598

The line in Table B used to select SRS is 188. The SRS of size 12 (labels) is 37, 08, 69, 58, 33, 55, 03, 26, 25, 17, 36, 29. Note that to obtain the last number we had to continue on line 189.

The actual times corresponding to these labels are 103, 66, 403, 178, 101, 156, 53, 92, 91, 81, 102, 99, and if we compute their mean we get

Sample mean $\bar{x}$ = 127.08

Following is the histogram with $\bar{x}$ marked on the horizontal axis.

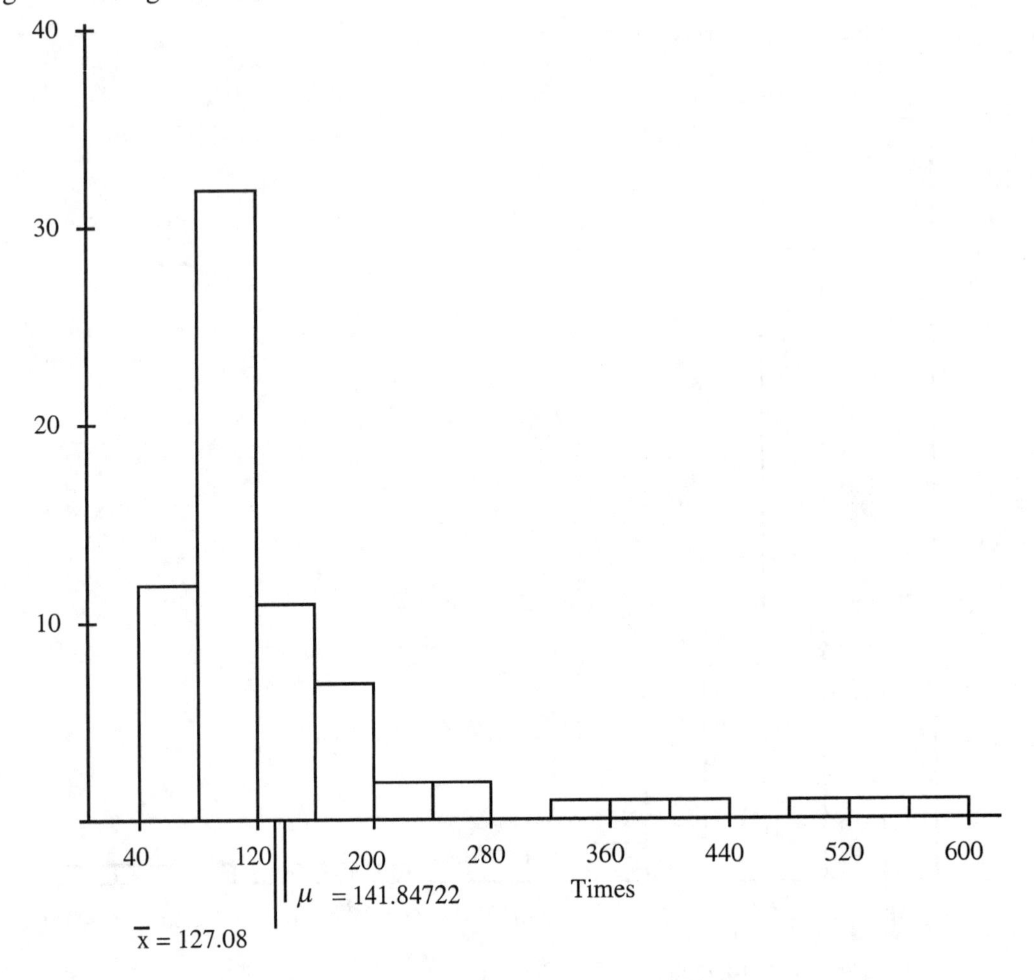

d) Here are the next four SRSs and the corresponding values of $\bar{x}$. The samples were selected beginning on line 189 where we left off in part (c). We continued on through line 192. For brevity, we list only the times corresponding to the labels selected followed by the mean of these times.

SRS 2 = 403, 162, 511, 92, 145, 81, 107, 126, 179, 99, 118, 128
$\bar{x}$ = 179.25

SRS 3 = 179, 88, 108, 81, 118, 102, 101, 380, 92, 56, 121, 53
$\bar{x}$ = 123.25

SRS 4 = 114, 108, 45, 137, 89, 101, 83, 162, 88, 179, 109, 128
$\bar{x}$ = 111.92

SRS 5 = 243, 107, 100, 329, 211, 45, 88, 73, 57, 83, 138, 118
$\bar{x}$ = 132.67

Here is the histogram with all five values of $\bar{x}$ marked on the horizontal axis.

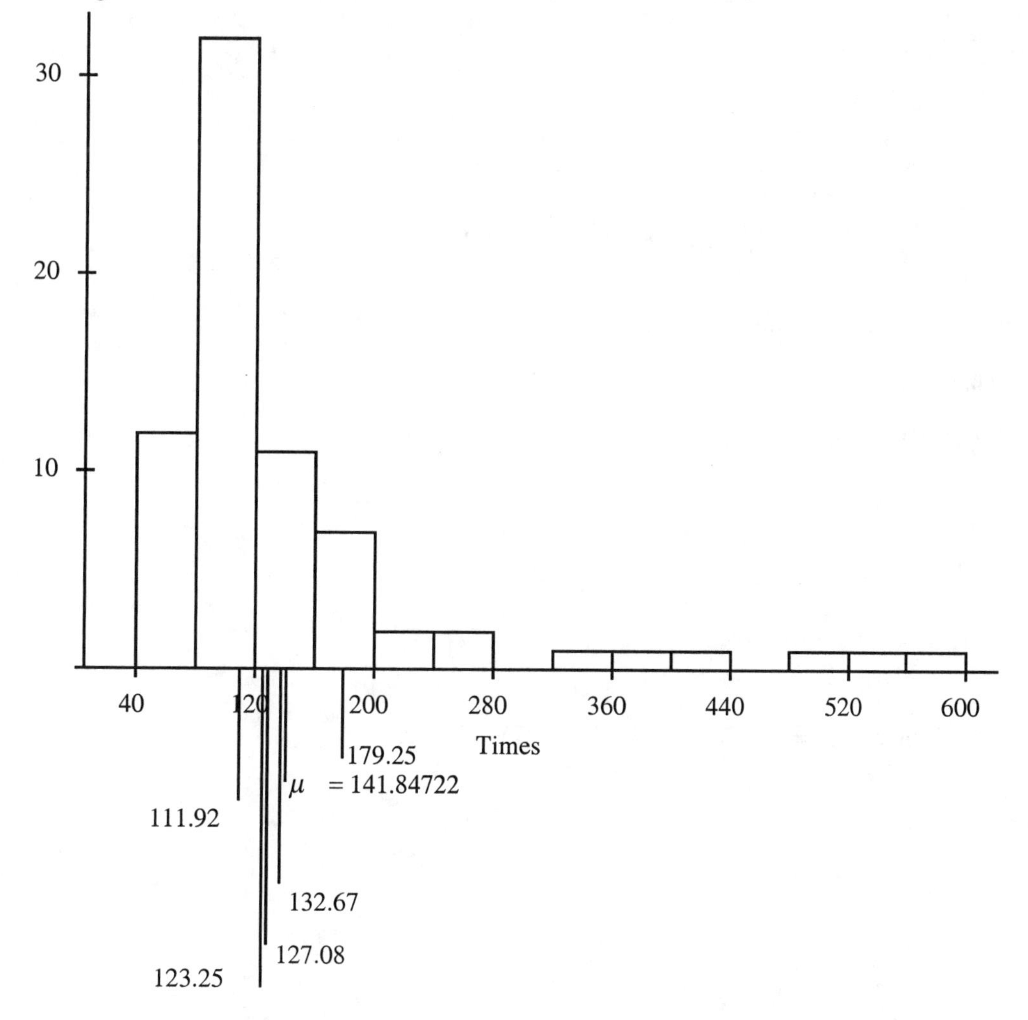

We would be somewhat surprised if all five $\bar{x}$'s fell on the same side of μ. The central limit theorem suggests that the sampling distribution of the $\bar{x}$'s should have a mean equal to μ and that the sampling distribution might be expected to look very roughly bell-shaped and symmetric. This would suggest that the probability that $\bar{x}$ would be above μ would be about 1/2, the same for the probability that $\bar{x}$ would be below μ. Thus getting all five $\bar{x}$'s on the same side of μ would be like flipping a coin five times and getting either all heads or all tails, a relatively uncommon occurrence.

e) As discussed in part (d), the central limit theorem suggests that the sampling distribution of the $\bar{x}$'s should have a mean equal to μ (notice that $\bar{x}$ is an unbiased estimator of μ) and that the sampling distribution might be expected to look very roughly bell-shaped and symmetric. This suggests that μ would be a reasonable measure of the center of this distribution.

Exercise 10.33

In this problem

$\mu = 4.22$
$\sigma = 0.127$
n = sample size = 5

From this information we find

$$\text{Center line} = \mu = 4.22$$

$$\text{Control limits} = \mu \pm 3\frac{\sigma}{\sqrt{n}} = 4.22 \pm 3\frac{0.127}{\sqrt{5}} = 4.22 \pm 0.17$$

so the lower control limit is 4.05 and the upper control limit is 4.39.

Exercise 10.35

KEY CONCEPTS - natural tolerances

We have

$\mu = 4.22$

$\sigma = 0.127$

and so

$$\text{Natural tolerances} = \mu \pm 3\sigma = 4.22 \pm 3(0.127) = 4.22 \pm 0.38$$

CHAPTER 11

GENERAL RULES OF PROBABILITY

OVERVIEW

Chapter 6 of your text discussed two-way tables and conditional distributions. In this section we learn about **conditional probabilities** and their use in calculating probabilities of complex events. The conditional probability of an event B given an event A has occurred is denoted $P(B|A)$ and is defined by

$$P(B|A) = \frac{P(A \text{ and } B)}{P(A)}$$

when $P(A) > 0$. In practice, a conditional probability can often be determined directly from the information given in a problem. Events are **independent** if knowledge that one event has occurred does not alter the probability that the second event occurs. Specifically, two events A and B are independent if $P(B|A) = P(B)$. From this, it follows that for independent events we must have the result that $P(A \text{ and } B) = P(A)P(B)$. In any particular problem we can use this result to check if two events are independent by seeing if the probabilities multiply correctly. However, most of the time independence is assumed as part of the probability model.

The following general rules are valid for any assignment of probabilities and allow us to compute the probabilities of events in many random phenomena:

Addition rule: If events A, B, C,... are all disjoint in pairs, then

$$P(\text{at least one of these events occur}) = P(A) + P(B) + P(C) + \cdots$$

Multiplication rule: If events A, B, C,... are independent, then

$$P(\text{all of the events occur}) = P(A)P(B)P(C)\cdots$$

General Addition Rule: For any two events A and B,

$$P(A \text{ or } B) = P(A) + P(B) - P(A \text{ and } B)$$

General multiplication rule: For any two events A and B,

$$P(A \text{ and } B) = P(A)P(B|A)$$

GUIDED SOLUTIONS

Exercise 11.3

KEY CONCEPTS - independence, multiplication rule

The key concept that must be properly understood to answer the questions raised in this problem is the notion of independence. Events A and B are independent if knowledge that A has occurred does not alter our assessment of the probability that B will occur.

What is required to use the multiplication rule to find the probability that A and B occur? Do you think the multiplication rule applies here?

Exercise 11.5

KEY CONCEPTS - independence, multiplication rule

What is the probability that a single light will remain bright for the holiday season? Now, what is the probability that all 20 will remain bright?

Exercise 11.7

KEY CONCEPTS - Venn diagrams

a) Let C be the event that a college student likes country music and G be the event that a college student likes gospel music. First write the three probabilities provided in terms of C and G and then add these results to the Venn diagram below. When you are done, your Venn diagram should look similar to the Venn diagram in Figure 11.5 of your text.

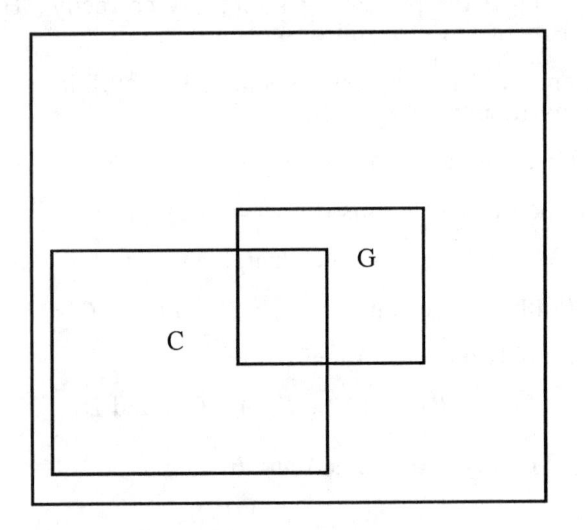

b) You should be able to read this probability easily off the Venn diagram. This event corresponds to the portion of C that doesn't overlap with G.

c) Students who like neither country nor gospel music are included in the region of the Venn diagram outside of the shapes corresponding to country and gospel.

Exercise 11.9

KEY CONCEPTS - two-way table of counts, conditional probabilities

The table from the problem is reproduced below to assist you.

	Bachelors	Master's	Professional	Doctorates	Total
Female	645	227	32	18	922
Male	505	161	40	26	732
Total	1150	388	72	44	1654

a) You can calculate this probability directly from the table. All degree recipients in the table are equally likely to be selected (that is what it means to select a degree recipient at random), so that the fraction of the degree recipients in the table that are women is the desired probability. How many women degree recipients are there? Where do you find this number in the table? What is the total number of degree recipients represented in the table? Use these numbers to compute the desired fraction.

b) This probability can also be calculated directly from the table. Since this is a conditional probability (i.e. this is a probability given that the degree recipient is a professional), we restrict ourselves only to professional degree recipients. The desired probability is then the fraction of these professional degree recipients that are women. Find the appropriate entries in the table to compute this fraction.

Exercise 11.13

KEY CONCEPTS - conditional probabilities, general multiplication rule

First define the events F = dollar will fall in value against the yen in the next month and R = supplier will demand renegotiation of the contract. You are given two probabilities in the exercise. One of the probabilities given is a conditional probability and these other is not. Write the two probabilities given in terms of the events F and R.

You are asked to find $P(F \text{ and } R)$. You should be able to evaluate this using the general multiplication rule and the information supplied in the problem.

$P(F \text{ and } R) =$

Exercise 11.25

KEY CONCEPTS - conditional probabilities, geometric probabilities

You want to use geometric arguments to evaluate

$$P(Y < 1/2 | Y > X) = \frac{P(Y < 1/2 \text{ and } Y > X)}{P(Y > X)}$$

In the previous equation, we have just used the definition of conditional probability where A corresponds to "$Y < 1/2$" and B corresponds to "$Y > X$." The three figures below are of the square $0 \le x \le 1$ and $0 \le y \le 1$. The figure on the left has the region corresponding to "$Y > X$" shaded. Remembering that probabilities correspond to areas, first evaluate $P(Y > X)$.

$P(Y > X) =$

The middle figure has the region corresponding to "$Y < 1/2$" shaded and the figure on the right has the region "$Y < 1/2$ and $Y > X$" shaded. The area of the shaded triangle in the figure on the right corresponds to the probability $P(Y < 1/2 \text{ and } Y > X)$. The value of this probability is

$P(Y < 1/2 \text{ and } Y > X) =$

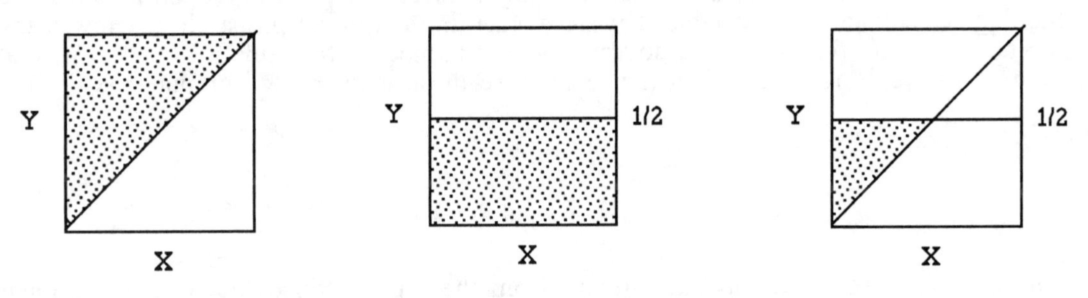

Putting this all together, you should be able to evaluate the probability given below.

$$P(Y < 1/2 | Y > X) = \frac{P(Y < 1/2 \text{ and } Y > X)}{P(Y > X)} =$$

Exercise 11.29

KEY CONCEPTS - conditional probabilities, sampling from large populations

a) What is the proportion of switches that are bad? This is the probability of drawing a bad switch because each switch has the same chance to be drawn.

P(switch drawn is bad) =

P(switch drawn is not bad) =

b) If the first switch drawn is bad, then for the second draw there are 9999 switches remaining of which 999 are bad. Use this information to calculate the conditional probability that the second switch is bad given that the first switch is bad.

P(Second switch drawn is bad | First switch drawn is bad) =

c) Now evaluate the conditional probability that the second switch is bad given that the first switch is not bad. Since this probability is almost the same as the probability in (b), we see that the trials are almost independent. The outcome of the first draw has little effect on the probabilities associated with the second draw.

P(Second switch drawn is bad | First switch drawn is not bad) =

Exercise 11.41

KEY CONCEPTS - addition rule, disjoint events

The following notation for events should prove helpful in this exercise and in Exercise 11.43 of this study guide. For the husband, you can use the events H_A, H_B, H_O and H_{AB} to denote the events that the husband has type A, type B, type O and type AB, respectively. What is the chance that Maria's husband has blood type B or type O?

$P(H_B \text{ or } H_O) =$

Exercise 11.43

KEY CONCEPTS - independence, addition rule, multiplication rule

In many problems, one of the greatest difficulties is deciding on some sort of notation which will allow you to express the problem simply. In this case, the following notation for events should prove helpful. For the husband, you can use the events H_A, H_B, H_O and H_{AB} to denote the events that the husband has type A, type B, type O and type AB, respectively. For the wife the notation for these events is W_A, W_B, W_O and W_{AB}. With these notational conventions, you can express more complicated events simply, and then apply the addition and multiplication rules learned in this chapter to compute the required probabilities.

a) The probability requested is $P(W_A \text{ and } H_B)$. Remember that blood types of married couples are independent, and the probabilities given for the blood types are valid for both men and women. What rule can be used to find the required probability?

$P(W_A \text{ and } H_B) =$

b) The probability requested is not the same as in part (a), although some of the calculations will be the same. In this part we are not told which spouse has type A blood and which spouse has type B blood. The probability requested is given below. See if you can apply the addition and multiplication rules correctly to get to the answer.

$P([W_A \text{ and } H_B] \text{ or } [H_A \text{ and } W_B]) =$

COMPLETE SOLUTIONS

Exercise 11.3

Suppose the events "college-educated" and "laborer or operator" are independent. This would imply that knowing whether someone was college-educated would not change the probability that they were a laborer or operator. In terms of this problem, if the events were independent, 14% of the entire labor force would work as laborers or operators, and also 14% of the college-educated labor force would work as laborers or operators (Knowledge of whether or not an individual has four years of college doesn't alter [increase or decrease] their chance of being a laborer or operator). The independence is what allows us to just multiply these probabilities together. The use of the formula (0.27)(0.14) = 0.038 to get the answer requires that 14% of the college students are laborers or operators. However, we would guess that fewer than 14% of those with four years of college were laborers or operators so that multiplying the two probabilities together is not the correct way to get the answer. The answer obtained by multiplying the two probabilities together is too large.

Exercise 11.5

An individual light will remain bright for the holiday season with probability 1 – 0.02 = 0.98. The whole string will remain bright with probability

$$P(\text{all 20 remain bright}) = (0.98)^{20} = 0.6676,$$

using the fact that the lights fail independently of each other.

Exercise 11.7

a) We are told that $P(C) = 0.4$, $P(G) = 0.3$, and $P(C \text{ and } G) = 0.10$. This information is reported in the Venn diagram. Since 40% of students like country and 10% like both country and gospel, we must have 0.4 – 0.1= 0.3 in the region corresponding to country and not gospel. Since 30% of students like gospel and 10% like both gospel and country, we must have 0.3 – 0.1= 0.2 in the region corresponding to gospel and not country. We have also entered 0.1 in the region corresponding to both country and gospel. Finally, the region outside of the shapes corresponding to country and gospel must have 40% of the students since the four regions in the Venn diagram are disjoint and make up the entire sample space.

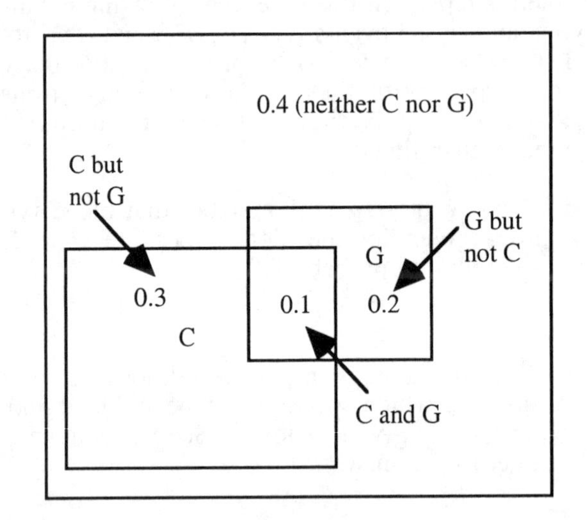

b) From the Venn diagram we see that 30% of students like country but not gospel.
c) From the Venn diagram we see that 40% of students like neither country nor gospel.

Exercise 11.9

a) The number of women degree recipients is found as the total for the first row and is (in thousands) 922. The total number of degree recipients in the table is in the lower right corner and is (in thousands) 1654. The desired probability is thus

(number of women degree recipients) / (total number of recipients in table) = 922/1654 = 0.5574

b) The desired conditional probability is

(number of professional degree recipients that are women) / (number of professional degree recipients)

= 32/72 = 0.4444

Exercise 11.13

You are given $P(F) = 0.4$ and the conditional probability $P(R|F) = 0.8$. Now you can use the general multiplication rule

$$P(F \text{ and } R) = P(F)P(R|F) = (0.4)(0.8) = 0.32.$$

The treasurer has assigned a personal probability of 32% that the dollar falls and the supplier demands renegotiation.

Exercise 11.25

KEY CONCEPTS - conditional probabilities, geometric probabilities

The figure on the left in the guided solution has the region corresponding to "$Y > X$" shaded, and because probabilities correspond to areas we see that $P(Y > X) = 1/2$. The figure on the right has the region "$Y < 1/2$ and $Y > X$" shaded. The area of the shaded triangle in the figure on the right is 1/8, so $P(Y < 1/2 \text{ and } Y > X) = 1/8$.

Putting this all together gives

$$P(Y < 1/2 | Y > X) = \frac{P(Y < 1/2 \text{ and } Y > X)}{P(Y > X)} = \frac{1/8}{1/2} = 0.25$$

Exercise 11.29

KEY CONCEPTS - conditional probabilities, sampling from large populations

a) $P(\text{switch drawn is bad}) = \frac{1000}{10,000} = 0.1$. $P(\text{switch drawn is not bad}) = 1 - 0.1 = 0.9$.

b) $P(\text{Second switch drawn is bad} \mid \text{First switch drawn is bad}) = \frac{999}{9999} = 0.09991$.

c) $P(\text{Second switch drawn is bad} \mid \text{First switch drawn is not bad}) = \frac{1000}{9999} = 0.10001$.

Exercise 11.41

The probability requested is $P(H_B \text{ or } H_O)$. The events H_B and H_O are disjoint (a person can't have two types of blood), so the addition rule for disjoint events applies.

$$P(H_B \text{ or } H_O) = P(H_B) + P(H_O) = 0.11 + 0.45 = 0.56.$$

Exercise 11.43

a) The probability requested is $P(W_A \text{ and } H_B)$. The events W_A and H_B are independent since the blood types of married couples are independent. (Note that these events are not disjoint. It's possible for both of them to occur together. When first learning about probability, students often confuse the ideas of independent and disjoint events). To evaluate the required probability we use the multiplication rule for independent events.

$$P(W_A \text{ and } H_B) = P(W_A)P(H_B) = 0.40 \times 0.11 = 0.044.$$

c) The probability requested is $P([W_A \text{ and } H_B] \text{ or } [H_A \text{ and } W_B])$, since, unlike part (a), the event doesn't specify which member of the couple has type A blood and which member has type B blood. The events $[W_A \text{ and } H_B]$ and $[H_A \text{ and } W_B]$ are disjoint, so to start we can write

$$P([W_A \text{ and } H_B] \text{ or } [H_A \text{ and } W_B]) = P([W_A \text{ and } H_B]) + P([H_A \text{ and } W_B]).$$

Each of the probabilities on the right hand side can be computed as in part (a), giving

$$P([W_A \text{ and } H_B] \text{ or } [H_A \text{ and } W_B]) = (0.40 \times 0.11) + (0.40 \times 0.11) = 0.088.$$

CHAPTER 12

BINOMIAL DISTRIBUTIONS

OVERVIEW

One of the most common situations giving rise to a **count** X is the **binomial setting**. The binomial setting consists of four assumptions about how the count was produced. They are

- the number n of observations is fixed
- the n observations are all independent
- each observation falls into one of two categories called "success" and "failure"
- the probability of success p is the same for each observation

When these assumptions are satisfied, the number of successes, X, has a **binomial distribution** with n trials and success probability p. For smaller values of n, the probabilities for X can be found easily using statistical software or the exact **binomial probability formula**. The formula is given by

$$P(X = k) = \binom{n}{k} p^k (1-p)^{n-k}$$

where $k = 0, 1, 2, ..., n$, and $\binom{n}{k} = \frac{n!}{k!(n-k)!}$ is called the **binomial coefficient**.

When the population is much larger than the sample, a count X of successes in an SRS of size n has approximately the binomial distribution with n equal to the sample size and p equal to the proportion of successes in the population.

The mean of a binomial random variable X is

$$\mu = np$$

and the standard deviation is

$$\sigma = \sqrt{np(1-p)}.$$

When n is large the count X is approximately $N(np, \sqrt{np(1-p)})$. This approximation should work well when $np \geq 10$ and $n(1 - p) \geq 10$.

GUIDED SOLUTIONS

Exercise 12.5

KEY CONCEPTS - binomial probabilities

a) For a binomial distribution with $n = 5$, what are the possible values of X.

b) X, the number of children who have type O blood, has the binomial distribution with $n = 5$ and $p = 0.25$. You need to find the probabilities of each of the values of X in part (a). The exact binomial probability formula is given by

$$P(X = k) = \binom{n}{k} p^k (1 - p)^{n-k}$$

and the required probabilites can be found by plugging the appropriate values of n, k and p in the formula. Alternatively, software can be used to evaluate the probabilities, although you should check one or two probabilities to make sure you understand how to use the formula. Fill in the values of the probabilities below.

$P(X = 0) =$

$P(X = 1) =$

$P(X = 2) =$

$P(X = 3) =$

$P(X = 4) =$

$P(X = 5) =$

Now draw a histogram of these probabilities. The heights of the bars should correspond to the probabilities. Typically the bars are centered at the possible values of X and are of width equal to 1. The first bar has been drawn for you.

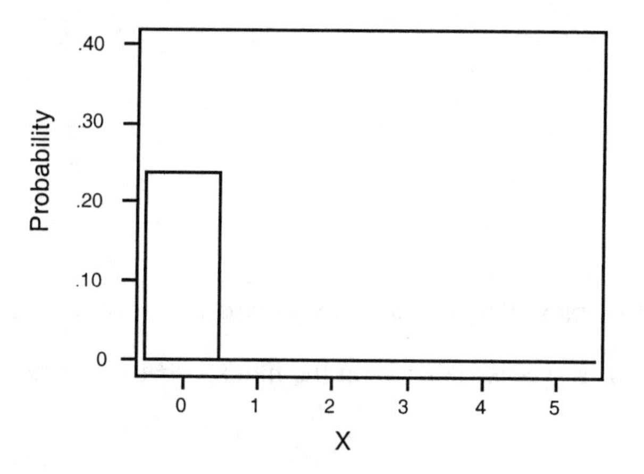

Exercise 12.13

KEY CONCEPTS - mean and standard deviation of a binomial count, normal approximation to the binomial distribution

a) When X is known to have a binomial distribution, you can use the formulas which express the mean and standard deviation of X in terms of n and p.

mean =

standard deviation =

b) This part can be done using the normal approximation for a count. First check to see that np and $n(1 - p)$ are both greater than or equal to 10. Then use the mean and standard deviation from (a) and the normal approximation to evaluate $P(X \leq 170)$.

np = ______________ and $n(1 - p)$ =__________________.

Can the approximation be used?

$P(X \leq 170) =$

Exercise 12.15

KEY CONCEPTS - binomial setting

There are four assumptions that need to be satisfied to ensure that the count *X* has a binomial distribution. The number of observations or trials must be fixed in advance and each trial must result in one of two outcomes. In each of the examples first try and identify both the number of trials and the two outcomes as these two assumptions are the most straightforward to check. Next, check that the trials are independent and the probability of success is the same from trial to trial. In each of the examples, do you think all four assumptions are satisfied? If not, try and explain why.

a)

b)

c)

Exercise 12.19

KEY CONCEPTS - binomial probabilities, mean and standard deviation of a binomial count

a) How many trials are there? If a success is a rise in the index, what is the success probability?

$n =$ $p =$

b) What are the possible values of *X*?

c) Use either the exact binomial formula or statistical software to evaluate the probabilities of each possible value of *X* and then draw a probability histogram for the distribution of *X* on the next page. You may want to refer to Exercise 12.5 of this study guide if you are having difficulties.

$P(X = 0) =$

$P(X = 1) =$

$P(X = 2) =$

$P(X = 3) =$

$P(X = 4) =$

$P(X = 5) =$

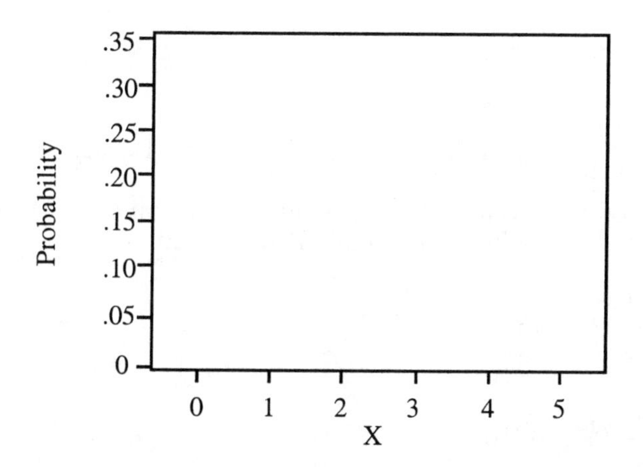

d) When X is known to have a binomial distribution, you can use the formulas which express the mean and standard deviation of X in terms of n and p. Make sure to mark the location of the mean on your histogram.

mean =

standard deviation =

Exercise 12.23

KEY CONCEPTS - binomial setting, mean of a binomial count, normal approximation to the binomial distribution

a) Review the four assumptions that need to be satisfied to ensure that the count X has a binomial distribution. Are they satisfied in this example? You may want to refer to Exercise 12.15of this study guide if you are having difficulty.

b) You can use the formulas which expresses the mean of X in terms of n and p.

$\mu =$

c) To use the normal approximation, you first need to check that np and $n(1-p)$ are both greater than or equal to 10. You then need to calculate the standard deviation of X. The mean has been computed in (b).

$\sigma =$

Using the mean and standard deviation that you have computed, you can now use the normal approximation to approximate $P(X \leq 70)$.

$P(X \leq 70) =$

d) If the mean computed in (b) is below 100, then you will need to take a larger sample to increase the mean number of respondents to 100. Using the formula for the mean, what value of n is required to make the value of μ equal to 100?

COMPLETE SOLUTIONS

Exercise 12.5

a) For a binomial distribution, the possible values of X are 0, 1, ..., n. Because n = 5 in this example, the possible values of X are 0, 1, 2, 3, 4, 5.
b) To calculate the probability of each value of X, we can use the binomial formula or statistical software. In this exercise, the use of the binomial formula is illustrated.

$$P(X=0)=\binom{5}{0}(.25)^0(.75)^5=\frac{5!}{0!5!}(0.2373) = 0.2373$$

$$P(X=1)=\binom{5}{1}(.25)^1(.75)^4=\frac{5!}{1!4!}(0.25)(.3164)=0.3955$$

$$P(X=2)=\binom{5}{2}(.25)^2(.75)^3=\frac{5!}{2!3!}(.0625)(.4219)=0.2637$$

$$P(X=3)=\binom{5}{3}(.25)^3(.75)^2=\frac{5!}{3!2!}(.0156)(.5625)=0.0879$$

$$P(X=4)=\binom{5}{4}(.25)^4(.75)^1=\frac{5!}{4!1!}(.0039)(.75)=0.0146$$

$$P(X=5)=\binom{5}{5}(.25)^5(.75)^0=\frac{5!}{5!0!}(.0010)=0.0010$$

The probability histogram is given below.

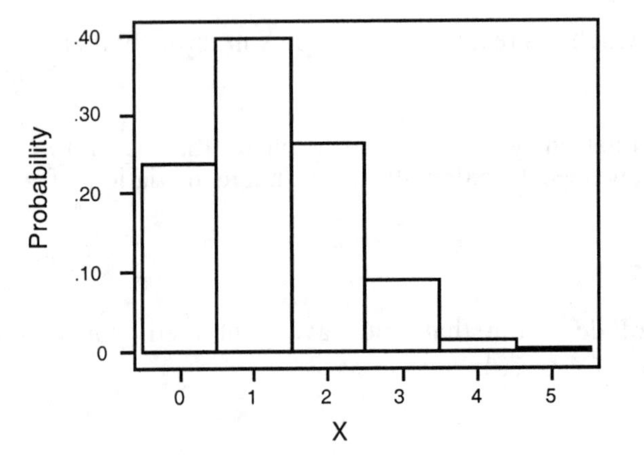

Exercise 12.13

a) X has a binomial distribution with n = 1500 and p = 0.12. The mean of X is np =1500(0.12) = 180, and the standard deviation of X is

$$\sqrt{np(1-p)}=\sqrt{1500(0.12)(0.88)}=12.586$$

b) The normal approximation can be used since both np = 180 and $n(1 - p)$ = 1320 are greater than 10. Using the mean and variance evaluated in (a) gives the approximation

$$P(X \leq 170) = P\left(\frac{X-180}{12.586} \leq \frac{170-180}{12.586}\right) = P(Z \leq -0.79) = 0.2148.$$

Exercise 12.15

a) There are n = 50 students and each student either passes or doesn't pass the exam. It is also reasonable to assume that the results for the 50 students are independent and each student has the same chance of passing.

b) There are n = 10 problems on the exam and each problem can be solved correctly or incorrectly. However, the probability of success (answering correctly) is likely to increase since the student receives instruction after incorrect answers.

c) There are n = 10 solubility tests and on each test the substance dissolves completely or it doesn't. However, temperature may affect the outcome of the test so the probability of dissolving is not necessarily the same on each test.

Exercise 12.19

a) X has a binomial distribution with n = 5 (the number of years to be observed) and p = 0.65 (the probability the index will increase in any given year). The independence of years is assumed as part of the model.

b) Because n = 5, the possible values are X are 0, 1, 2, 3, 4, 5.

c) To calculate the probability of each value of X, we can use the binomial formula or statistical software This is very similar to Exercise 12.5 of this study guide in which the use of the binomial formula was illustrated. The only difference is that p = 0.65 in this exercise and p was 0.25 in Exercise 12.5. The probabilities listed below were obtained using the Minitab software.

```
Binomial with n = 5 and p = 0.650000
        x              P(X = x)
      0.00              0.0053
      1.00              0.0488
      2.00              0.1811
      3.00              0.3364
      4.00              0.3124
      5.00              0.1160
```

The probability histogram corresponding to this distribution is given below.

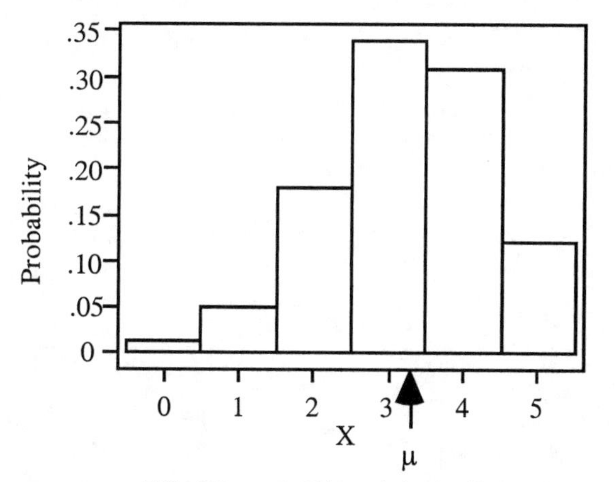

d) The mean of X is $\mu = np$ = 5(0.65) = 3.25 and is indicated on the histogram in part (c). The standard deviation of X is

$$\sigma = \sqrt{np(1-p)} = \sqrt{5(0.65)(0.35)} = 1.067$$

Exercise 12.23

a) The Binomial distribution with n observations and probability p of success gives a good approximation to the sampling distribution of the count of successes in an SRS of size n from a large population containing proportion p of successes. In this case, we believe p in the population (businesses listed in the Yellow Pages) is about 0.5 and the number of observations (size of sample) is 150. The Binomial distribution with $n = 150$ and $p = 0.5$ is reasonable.

b) The mean of the binomial is $\mu = np = 150(0.50) = 75$, so we expect 75 businesses to respond.

c) To apply the normal approximation we first check that $np = 150(0.5) = 75$ and $n(1 - p) = 100(0.5) = 75$ are both greater than 10, which they are. The next step is to evaluate the mean and standard deviation of the binomial distribution. From part (b), $\mu = np = 100(0.75) = 75$, and the standard deviation of X is

$$\sigma = \sqrt{np(1-p)} = \sqrt{150(0.5)(0.5)} = 6.12.$$

Next, we act as though X had the $N(75, 6.12)$ distribution.

$$\begin{aligned} P(X \le 70) &= P\left(\frac{X-75}{6.12} \le \frac{70-75}{6.12}\right) \\ &= P(Z \le -0.82) \\ &= 0.2061. \end{aligned}$$

There is about a 20% chance that fewer than 70 businesses will respond.

d) The mean is $\mu = np = n(0.50)$. To increase the mean number of respondents to 100 requires that n be increased to 200.

CHAPTER 13

CONFIDENCE INTERVALS: THE BASICS

OVERVIEW

A **confidence interval** provides an estimate of an unknown parameter of a population or process, along with an indication of how accurate this estimate is and how confident we are that the interval is correct. Confidence intervals have two parts. One is an interval computed from our data. This interval typically has the form

$$\text{estimate} \pm \text{margin of error}$$

The other part is the **confidence level**, which states the probability that the *method* used to construct the interval will give a correct answer. For example, if you use a 95% confidence interval repeatedly, in the long run 95% of the intervals you construct will contain the correct parameter value. Of course, when you apply the method only once, you do not know if your interval gives a correct value or not. Confidence refers to the probability that the method gives a correct answer in repeated use, not the correctness of any particular interval we compute from data.

Suppose we wish to estimate the unknown mean μ of a normal population with known standard deviation σ based on an SRS of size n. A level C confidence interval for μ is

$$\bar{x} \pm z^* \frac{\sigma}{\sqrt{n}}$$

where z^* is such that the probability is C that a standard normal random variable lies between $-z^*$ and z^* and is obtained from the bottom row in Table C. These z-values are called critical values.

The margin of error $z^* \frac{\sigma}{\sqrt{n}}$ of a confidence interval decreases when any of the following occur:

- The confidence level C decreases.
- The sample size n increases.
- The population standard deviation σ decreases.

The sample size needed to obtain a confidence interval for a normal mean of the form

$$\text{estimate} \pm \text{margin of error}$$

with a specified margin of error m is

$$n = \left(\frac{z^* \sigma}{m}\right)^2$$

where z^* is the critical value for the desired level of confidence. Many times, the n you will find will not be an integer. If it is not, round up to the next larger integer.

The formula for any specific confidence interval is a recipe that is correct under specific conditions. The most important conditions concern the methods used to produce the data. Many methods (including those discussed in this section) assume that our data were collected by random sampling. Other conditions, such as the actual distribution of the population, are also important.

GUIDED SOLUTIONS

Exercise 13.1

KEY CONCEPTS - confidence intervals, interpreting statistical confidence

a) The general form of a confidence interval is

$$\text{estimate} \pm \text{margin of error}$$

Identify the estimate (the percentage of people who said "Yes" when asked, "Would you like to lose weight?") and the margin of error. Then combine these in the general form of a confidence interval as indicated.

b) In formulating your explanation, consider the meaning of statistical confidence as described in Chapter 13 of your textbook or in the overview for this chapter of the Study Guide. Write your explanation.

Exercise 13.9

KEY CONCEPTS - confidence intervals for means, the effect of sample size on the margin of error

a) First identify the following quantities. You will want to use the bottom row of Table C to find z^*.

$\bar{x}$ =

σ =

n =

z^* (for a 95% confidence interval) =

Now use the formula for a 95% confidence interval to compute the desired interval

$$\bar{x} \pm z^* \frac{\sigma}{\sqrt{n}} =$$

b) Use the same formula as in part (a), but now with $n = 250$ rather than 1000.

$$\bar{x} \pm z^* \frac{\sigma}{\sqrt{n}} =$$

c) Once again, use the same formula as in part (a), but now with $n = 4000$.

$$\bar{x} \pm z^* \frac{\sigma}{\sqrt{n}} =$$

d) The margins of errors are the quantities after the $\pm$. You computed these in parts (a), (b), and (c). List them here.

Margin of error for $n = 250$:

Margin of error for $n = 1000$:

Margin of error for $n = 4000$:

What pattern do you observe?

Exercise 13.14

KEY CONCEPTS - confidence intervals for means, the effect of changing the confidence level

a) Make your stemplot with split stems in the space provided below. Refer to Chapter 1 of your textbook if you need to refresh your memory about stemplots.

1 |
1 |
2 |
2 |
3 |
3 |
4 |

Are there any outliers? Is there extreme skewness?

b) Identify σ, n, $\bar{x}$ (you will need to calculate this), and z^* for a 90% confidence interval (use Table C). Then use the formula to calculate the confidence interval.

$$\bar{x} \pm z^* \frac{\sigma}{\sqrt{n}} =$$

c) What is the relation between the level of confidence and the width of the interval? Why is this relation not surprising?

Exercise 13.21

KEY CONCEPTS - sample size required to obtain a confidence interval of specified margin of error

Refer to Exercise 13.14 part (b) of this Study Guide for the appropriate values of σ and z^*. What margin of error m is desired? Now complete the following to compute the necessary sample size n.

$$n = \left(\frac{z^*\sigma}{m}\right)^2 =$$

Remember to round up to the nearest integer for your final answer.

Exercise 13.23

KEY CONCEPTS - the effect of sample size on the margin of error

Is the number of respondents aged 18 to 29 years, larger than, smaller than, or equal to 1002? What effect does a change in sample size have on the margin of error?

Exercise 13.29

KEY CONCEPTS - interpreting confidence intervals

Review the meaning of statistical confidence. In the long run, what is supposed to be within three percentage points of at least 95% of the results?

Complete Solutions

Exercise 13.1

a) The estimate of interest here (the percent of people who said "Yes" when asked, "Would you like to lose weight?") is 51%. Since the margin of error for a 95% confidence interval is ± 3%, the 95% confidence interval for the percent of all adult women who think they do not get enough time for themselves is 51% ± 3%, or between 48% and 54%.

b) Suppose we take all possible random samples of national adults. In each sample, suppose we determine the percent of people who say "Yes" when asked, "Would you like to lose weight?" For each of these percents, suppose we add and subtract the margin of error for a 95% confidence interval. Of the resulting intervals, 95% will contain the actual percent of all adults in the United States who would like to lose weight. This is what we mean by "95% confidence." Note that we do not know if any particular interval (such as the 48% to 54% interval in part (a) contains the true value of the percent. The confidence level of 95% refers only to the percent of the intervals produced by all samples that will contain the true percent.

Exercise 13.9

a) We are given that the sample mean $\bar{x}$ is 22, the standard deviation σ is 50, and the sample size n is 1000. For 95% confidence, $z^* = 1.96$. Thus a 95% confidence interval for the mean score μ in the population of all young women is

$$\bar{x} \pm z^* \frac{\sigma}{\sqrt{n}} = 22 \pm 1.96 \frac{50}{\sqrt{1000}} = 22 \pm 3.10$$

or 18.90 to 225.10.

b) We simply replace $n = 1000$ by $n = 250$ in our calculations and get

$$\bar{x} \pm z^* \frac{\sigma}{\sqrt{n}} = 22 \pm 1.96 \frac{50}{\sqrt{250}} = 22 \pm 6.20$$

or 15.80 to 28.20.

c) We use $n = 4000$ in our calculations and get

$$\bar{x} \pm z^* \frac{\sigma}{\sqrt{n}} = 22 \pm 1.96 \frac{50}{\sqrt{4000}} = 22 \pm 1.55$$

or 19.45 to 23.55.

d) Margin of error for $n = 250$: ± 6.20

Margin of error for $n = 1000$: ± 3.10

Margin of error for $n = 4000$: ± 1.55

We see that as the sample size increases, the margin of error decreases.

Exercise 13.14

a) Here is a stemplot of the data. We have used split stems. There are no outliers. There does seem to be some slight right skewness, but it is certainly not extreme.

```
1|124
1|8
2|2233
2|6789
3|034
3|55
4|0
```

b) We are told that $\sigma = 8$ and we know that $n = 18$. From the data we calculate $\bar{x} = 25.67$. From Table C we find $z^* = 1.645$ for a 90% confidence interval. Thus our 90% confidence interval is

$$\bar{x} \pm z^* \frac{\sigma}{\sqrt{n}} = 25.67 \pm 1.645 \frac{8}{\sqrt{18}} = 25.67 \pm 3.10$$

or (22.57, 28.77).

c) Her interval will be wider. It is not surprising that the confidence interval becomes wider as you increase the confidence level. One would expect to be more confident that your interval includes the population mean as you increase the width of the interval.

Exercise 13.21

We want a 90% confidence interval with a margin of error $m = 1$. As we saw in part (b) of Exercise 13.14 of this Study Guide, $\sigma = 8$, and from Table C we find $z^* = 1.645$ for a 90% confidence interval. The formula for the proper sample size n is

$$n = \left(\frac{z^* \sigma}{m}\right)^2 = \left(\frac{1.645 \times 8}{1}\right)^2 = 173.2$$

which we round up to $n = 174$.

Exercise 13.23

The poll consists of 1002 adults. Respondents aged 18 to 29 years are only a portion of all respondents, and the number of respondents aged 18 to 29 years must be less than 1002. Recall that the margin of error increases as sample size increases. Thus, the margin of error for respondents aged 18 to 29 years must be larger than ± 3 percentage points.

Exercise 13.29

If you use a 95% confidence interval repeatedly, in the long run 95% of the intervals you construct will contain the correct *parameter value* (in this case the percentage of all adults), not the result of this particular survey. Thus, the last sentence in the Associated Press quote should be "This means that, if the same questions were repeated in 20 polls, the results of at least 19 surveys would be within three percentage points of the percentage of *all adults* in the population."

CHAPTER 14

TESTS OF SIGNIFICANCE: THE BASICS

OVERVIEW

Tests of significance and confidence intervals are the two most widely used types of formal statistical inference. A test of significance is done to assess the evidence against the **null hypothesis** H_0 in favor of an **alternative hypothesis** H_a. Typically the alternative hypothesis is the effect that the researcher is trying to demonstrate, and the null hypothesis is a statement that the effect is not present. The alternative hypothesis can be either **one-** or **two-sided**.

Tests are usually carried out by first computing a **test statistic**. The test statistic is used to compute a ***P*-value**, which is the probability of getting a test statistic at least as extreme as the one observed, where the probability is computed when the null hypothesis is true. The *P*-value provides a measure of how incompatible our data are with the null hypothesis, or how unusual it would be to get data like ours if the null hypothesis were true. Since small *P*-values indicate data that are unusual or difficult to explain under the null hypothesis, we typically reject the null hypothesis in these cases. In this case, the alternative hypothesis provides a better explanation for our data.

Significance tests of the null hypothesis H_0: $\mu = \mu_0$ with either a one- or a two-sided alternative are based on the test statistic

$$z = \frac{\bar{x} - \mu_0}{\sigma / \sqrt{n}}$$

The use of this test statistic assumes that we have an SRS from a normal population with known standard deviation σ. When the sample size is large, the assumption of normality is less critical because the sampling distribution of $\bar{x}$ is approximately normal. *P*-values for the test based on z are computed using Table A.

When the *P*-value is below a specified value α, we say the results are **statistically significant at level α,** or we reject the null hypothesis at level α. Tests can be carried out at a fixed significance level by obtaining the appropriate critical value z^* from the bottom row in Table C.

GUIDED SOLUTIONS

Exercise 14.2

KEY CONCEPTS - testing hypotheses about means

a) If the null hypothesis H_0: $\mu = 115$ is true, then scores in the population of older students are normally distributed, with mean $\mu = 115$ and standard deviation $\sigma = 30$. What then is the sampling distribution of $\bar{x}$, the mean of a sample of size $n = 25$? (We studied the sampling distribution of $\bar{x}$ in Chapter 10 of your textbook)

Sketch the density curve of this distribution. Be sure to label the horizontal axis properly.

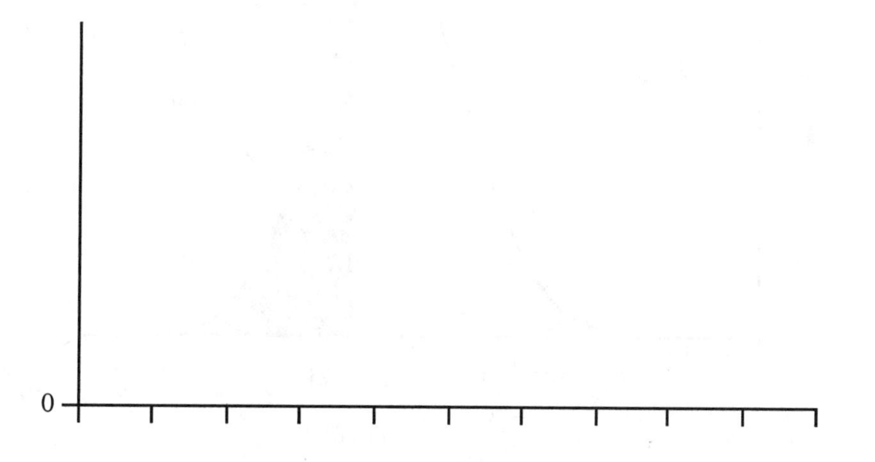

b) Mark the two points on your sketch in part (a). Referring to this sketch, explain in simple language why one result is good evidence that the mean score of all older students is greater than 115 and why the other outcome is not. Think about how far out on the density curve the two points are.

Exercise 14.9

KEY CONCEPTS - testing hypotheses about means, the z test statistic

We are testing the hypothesis H_0: $\mu = 115$, so $\mu_0 = 115$. The two outcomes in Exercise 14.2 are $\bar{x} = 118.6$ and $\bar{x} = 125.8$. The population standard deviation is $\sigma = 30$, and the sample size is $n = 25$. To compute the values of the test statistic z, complete the following.

For $\bar{x} = 118.6$: $z = \dfrac{\bar{x} - \mu_0}{\sigma / \sqrt{n}} =$

For $\bar{x} = 125.8$: $z = \dfrac{\bar{x} - \mu_0}{\sigma / \sqrt{n}} =$

Exercise 14.12

KEY CONCEPTS - P-values, statistical significance

a) Refer to the graph on the next page. The P-value for 118.6 is the shaded area, i.e., the area under the normal curve to the right of 118.6. We learned how to calculate such areas in Chapter 3 of your textbook. First you will need to find the z score of 118.6 and then you will need to use Table A to find the area to the right of this z score under a standard normal curve. Use the space for your calculations.

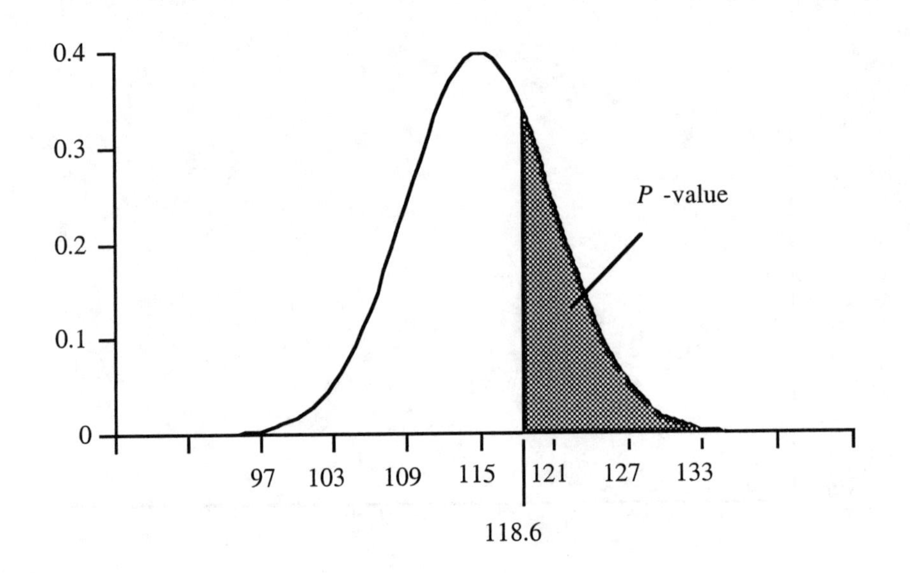

Now perform a similar calculation to find the *P*-value of 125.8.

Why do these values tell us that one outcome is strong evidence against the null hypothesis and that the other is not?

Exercise 14.15

KEY CONCEPTS - *P*-values, statistical significance

To answer this question, recall that an observed value is statistically significant at level α if the *P*-value is smaller than α. Use this fact and your solutions to Exercise 14.12 in this Study Guide to answer the question.

Exercise 14.22

KEY CONCEPTS - testing hypotheses at a fixed significance level

a) The z-test statistic for testing against a two-sided alternative, as in this problem, is $|z| = \left|\frac{\bar{x} - \mu_0}{\sigma / \sqrt{n}}\right|$. Identify μ_0, the standard deviation σ, the sample mean $\bar{x}$ and the sample size n. Then complete the test.

$$|z| = \left|\frac{\bar{x} - \mu_0}{\sigma / \sqrt{n}}\right| =$$

b) To answer you will have to find the appropriate critical value from Table C. Note that we are testing against a two-sided alternative.

c) Follow the procedure in part (b), but with significance level 1%.

d) Between what two adjacent critical values in Table C does your z-test statistic lie? What are the corresponding tail probabilities at the top of the table? What do you need to do to the tail probabilities to convert them to critical values for testing against a two-sided alternative?

Exercise 14.25

KEY CONCEPTS - relationship between two-sided tests and confidence intervals

a) The 95% confidence interval 31.5 ± 3.5 is equivalent to the interval 28 to 35. What must be true of the relationship between a $1-\alpha$ confidence interval for μ and the value μ_0 for a level α two-sided significance test of H_0: $\mu = \mu_0$ to reject the null hypothesis? Is this true here?

b) Is the condition mentioned in part (a) true here?

Exercise 14.45

KEY CONCEPTS - interpreting *P*-values

Recall that the *P*-value is the probability, computed supposing H_0 to be true, that the test statistic will take a value at least as extreme as that actually observed. Use this to evaluate the student's statement.

COMPLETE SOLUTIONS

Exercise 14.2

a) From Chapter 10 we know that the sampling distribution of $\bar{x}$ is normal with mean μ = 115 and standard deviation $\sigma = 30/\sqrt{n} = 30/\sqrt{25} = 30/5 = 6$. A sketch of the density curve of this distribution follows.

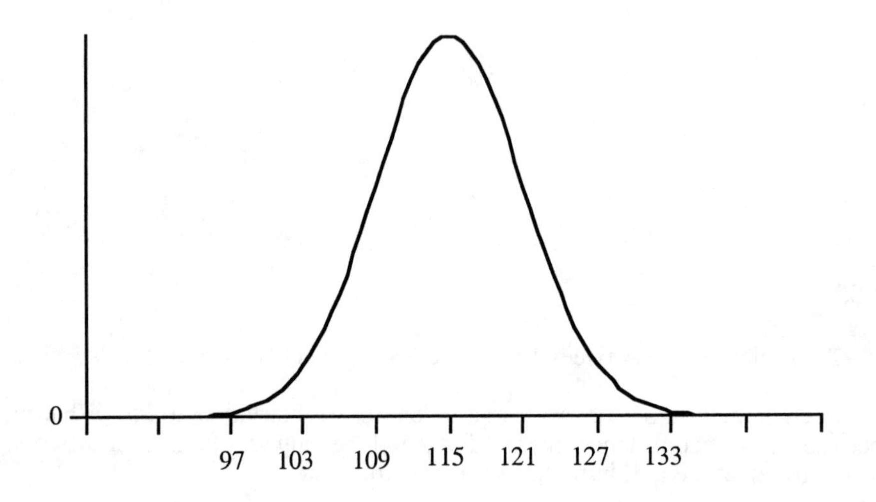

b) The two points are marked on the following curve.

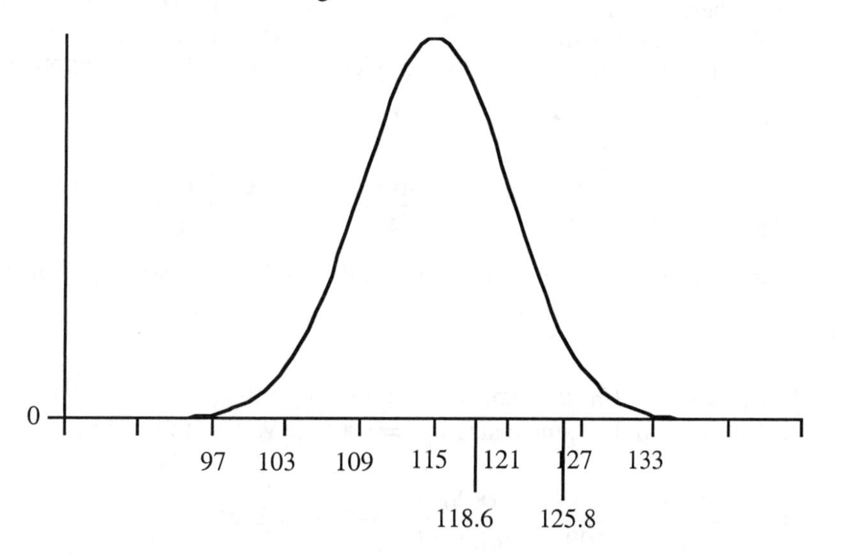

The 125.8 is much farther out on the normal curve than 118.6. In other words, it would be unlikely to observe a mean of 125.7 if the null hypothesis H_0: $\mu = 115$ is true. However, a mean of 118.6 is fairly likely if the null hypothesis is true. A mean as large as 125.8 is more likely to occur if the true mean is larger than 115. Thus, 125.8 is good evidence that the mean score of all older students is greater than 115, while a mean score of 118.6 is not.

Exercise 14.9

For $\bar{x} = 118.6$: $z = \dfrac{\bar{x} - \mu_0}{\sigma / \sqrt{n}} = \dfrac{118.6 - 115}{30 / \sqrt{25}} = \dfrac{3.6}{6} = 0.6.$

For $\bar{x} = 125.8$: $z = \dfrac{\bar{x} - \mu_0}{\sigma / \sqrt{n}} = \dfrac{125.8 - 115}{30 / \sqrt{25}} = \dfrac{10.8}{6} = 1.8$

Exercise 14.12

a) Refer to the graph in the guided solution. The *P*-value for 118.6 is the shaded area, i.e., the area under the normal curve to the right of 118.6. We learned how to calculate such areas in Chapter 3 of your textbook. First we compute the z score of 118.6. We did this in Exercise 14.9 of this Study Guide and found

$$z \text{ score} = \frac{\bar{x} - \mu_0}{\sigma / \sqrt{n}} = \frac{118.6 - 115}{30 / \sqrt{25}} = \frac{3.6}{6} = 0.6$$

From Table A we find that the area under the standard normal curve to the right of 0.6 is 1 minus the area to the left of 0.6 = 1 – 0.7527 = 0.2473.

We make a similar calculation for the *P*-value of 125.8:

$$z \text{ score} = \frac{\bar{x} - \mu_0}{\sigma / \sqrt{n}} = \frac{125.8 - 115}{30 / \sqrt{25}} = \frac{10.8}{6} = 1.8$$

The area to the right of this z score under a standard normal curve = 1 minus the area to the left of 1.80 = 1 – 0.9641 = 0.0359.

In summary,

P-value of 118.6 = 0.2473

P-value of 125.8 = 0.0359

The P-value for 118.6 is not particularly small and so this outcome is reasonably likely to happen by chance if the null hypothesis is true. Thus, the outcome 118.6 is not strong evidence against the null hypothesis. The P-value for 125.8 is small and so this outcome is unlikely to happen by chance if the null hypothesis is true. Thus, the outcome 125.8 is strong evidence against the null hypothesis.

Exercise 14.15

The P-value for 125.8 is less than 0.05, so it is statistically significant at the $\alpha = 0.05$ level. The P-value for 118.6 is larger than 0.05, so it is not statistically significant at the 0.05 level.

Neither P-value is less than 0.01, so neither observed value would be statistically significant at $\alpha = 0.01$.

Exercise 14.22

a) Because the null hypothesis is H_0: $\mu = 0.5$, we have $\mu_0 = 0.5$; the standard deviation is $\sigma = 0.2887$, the sample mean is $\bar{x} = 0.4365$, and the sample size is $n = 100$; the value of the z-test statistic is

$$|z| = \left|\frac{\bar{x} - \mu_0}{\sigma / \sqrt{n}}\right| = \left|\frac{0.4365 - 0.5}{0.2887 / \sqrt{100}}\right| = \left|\frac{-0.0635}{0.02887}\right| = |-2.20| = 2.20$$

b) Because we are testing against a two-sided alternative, we need to find the upper $\alpha/2 = 0.05/2 = 0.025$ critical value in Table C. We see that this critical value is 1.96. Since our z-test statistic is greater than this critical value, the result is significant at the 0.05 level.

c) Now we look for the upper $\alpha/2 = 0.01/2 = 0.005$ critical value in Table C. We see that this critical value is 2.576. Since our z-test statistic is smaller than this critical value, the result is not significant at the 0.01 level.

d) As we examine the critical values in Table C, we observe that our z-test statistic of 2.20 lies between the critical values 2.054 and 2.326. The corresponding upper tail probabilities at the top of the table are 0.02 and 0.01. Since we are testing against a two-sided alternative, we must multiply the tail probabilities by 2 to get the appropriate critical values (we double the values because we need to include both the upper and lower tail probabilities). Therefore we see that 2.20 lies between the 0.04 and 0.02. We conclude that $0.02 < P\text{-value} < 0.04$.

Exercise 14.25

a) The null hypothesis is H_0: $\mu = 34$. The value of μ under H_0, namely 34, falls inside the 95% confidence interval and we would not reject H_0.

b) The null hypothesis is H_0: $\mu = 36$. The value of μ under H_0, namely 36, falls outside the 95% confidence interval and we would reject H_0.

Exercise 14.45

The student's statement is not correct. The null hypothesis is either true or false; statements about the probability that it is true are not meaningful. The P-value is the probability, computed supposing H_0 to be true, that the test statistic will take a value at least as extreme as that actually observed. Thus, P-values tell us about how strong our data are as evidence against the null hypotheses. Small values indicate strong evidence against the null hypothesis.

CHAPTER 15

INFERENCE IN PRACTICE

OVERVIEW

Statistical inference from data based on a badly designed survey or experiment is often useless. Remember, a statistical test is valid only under certain conditions with data that have been properly produced. Whenever you use statistical inference, you are assuming your data are a probability sample or come from a randomized comparative experiment.

Always do data analysis before inference to detect outliers or other problems that would make your inference untrustworthy.

The margin of error of a confidence interval only accounts for the chance variation due to random sampling. Errors due to nonresponse or undercoverage are often more serious in practice.

When describing the outcome of a hypothesis test, it is more informative to give the P-value than to just reject or not reject a decision at a particular significance level α. The traditional levels of 0.01, 0.05 and 0.10 are arbitrary and serve as rough guidelines. Different people will insist on different levels of significance depending on the plausibility of the null hypothesis and the consequences of rejecting the null hypothesis. There is no sharp boundary between significant and insignificant, only increasingly strong evidence as the P-value decreases.

When testing hypotheses with a very large sample, the P-value can be very small for effects that may not be of interest. Don't confuse small P-values with large or important effects. Statistical significance is not the same as practical significance. Plot the data to display the effect you are trying to show and also give a confidence interval that says something about the size of the effect.

Just because a test is not statistically significant doesn't imply that the null hypothesis is true. Statistical significance may occur when the test is based on a small sample size. Finally, if you run enough tests, you will invariably find statistical significance for one of them. Be careful in interpreting the results when testing many hypotheses on the same data.

From the point of view of making decisions, H_0 and H_a are just two statements of equal status that we must decide between. One chooses a rule for deciding between H_0 and H_a on the basis of the probabilities of the two types of errors we can make. A **Type I error** occurs if H_0 is rejected when it is in fact true. A **Type II error** occurs if H_0 is accepted when in fact H_a is true. There is a clear relation between α-level significance tests and testing from the decision-making point of view. α is the probability of a Type I error.

To compute the Type II error probability of a significance test about a mean of a normal population:

- Write the rule for accepting the null hypothesis in terms of $\bar{x}$.

- Calculate the probability of accepting the null hypothesis when the alternative is true.

The **power** of a significance test is always calculated at a specific alternative hypothesis and is the probability that the test will reject H_0 when that alternative is true. The power of a test against any particular alternative is 1 minus the probability of a Type II error. Power is usually interpreted as the ability of a test to detect an alternative hypothesis or as the sensitivity of a test to an alternative hypothesis. The power of a test can be increased by increasing the sample size when the significance level remains fixed.

GUIDED SOLUTIONS

Exercise 15.3

KEY CONCEPTS - sources of error and confidence intervals

To answer these questions, recall that the margin of error of a confidence interval only accounts for the chance variation due to random sampling. Errors due to nonresponse or undercoverage are often more serious in practice.

a) Is this source of error included in the margin of error? Yes _______ No _______

b) Is this source of error included in the margin of error? Yes _______ No _______

c) Is this source of error included in the margin of error? Yes _______ No _______

Exercise 15.6

KEY CONCEPTS - statistical significance versus practical importance

In this problem we see that the *P*-value associated with the outcome $\bar{x} = 478$ depends on the sample size. The probability of getting a value of $\bar{x}$ as large as 478 if the mean is 475 will become smaller as the sample size gets larger. (Do you remember why? Look at the formula for the z score of $\bar{x}$.) Since this probability is the *P*-value, we see that a small effect is more likely to be detected for larger sample sizes than for smaller sample sizes. But this doesn't necessarily make the effect interesting or important. A confidence interval tells you something about the size of the effect, not the *P*-value.

a) Find the *P*-value by computing the z-test statistic and the probability of exceeding it.

b) This is the same as part (a) but with a larger sample size. The larger sample size makes the probability of getting a value of $\bar{x}$ as large as 478 smaller than it was in part (a). Compute the *P*-value.

c) Find the *P*-value in this last case. It will be the smallest. Why?

Exercise 15.9

KEY CONCEPTS - multiple analyses

a) What does a *P*-value less than 0.01 mean? Out of 500 subjects, how many would you expect to achieve a score that has such a *P*-value if all 500 are guessing?

b) What would you suggest the researcher now do to test whether any of these four subjects have ESP?

Exercise 15.11

KEY CONCEPTS - power

Begin by rewriting the rule for rejecting H_0 in terms of $\bar{x}$. We help you by getting you started. The rule is to reject H_0 if $z \leq -1.645$, or

$$z = \frac{\bar{x} - 300}{3/\sqrt{6}} = \frac{\bar{x} - 300}{1.22} \leq -1.645.$$

What does this inequality imply about the values of $\bar{x}$?

a) If $\mu = 299$, what is the sampling distribution of $\bar{x}$? Use this to compute the probability that the test rejects, i.e., the probability that $\bar{x}$ takes on values (which you computed above) that lead to rejecting H_0 when the particular alternative $\mu = 299$ is true.

b) If $\mu = 295$, what is the sampling distribution of $\bar{x}$? Use this to compute the probability that the test rejects, i.e., the probability that $\bar{x}$ takes on values (which you computed above) that lead to rejecting H_0 when the particular alternative $\mu = 295$ is true.

c) What do you notice in parts (a) and (b) about the change in power as the true value of μ changes from 299 to 295?

Exercise 15.13

KEY CONCEPTS - Type I and Type II error probabilities

a) Write the two hypotheses. Remember, we usually take the null hypothesis to be the statement of "no effect."

H_0:

H_a:

Describe the two types of errors as "false positive" and "false negative" test results.

b) Which error probability would you choose to make smaller (at the expense of making the other error probability larger) and why?

Exercise 15.15

KEY CONCEPTS - Type I and Type II error probabilities

a) If $\mu = 0$, what is the sampling distribution of $\bar{x}$? Now use this to compute the probability the test rejects, i.e., the probability $\bar{x} > 0$.

b) If $\mu = 0.3$, what is the sampling distribution of $\bar{x}$? Now use this to compute the probability the test accepts H_0, i.e., the probability $\bar{x} \leq 0$.

c) If $\mu = 1$, what is the sampling distribution of $\bar{x}$? Now use this to compute the probability the test accepts H_0, i.e., the probability $\bar{x} \leq 0$.

Exercise 15.33

KEY CONCEPTS - power and the relationship with the Type II error probability

What is the relationship between the probability of a Type I error and the level of significance?

What is the relationship between the power of a test at a particular alternative and the Type II error at this alternative? Now use the value of the power that you calculated in Exercise 15.11 of this Study Guide to compute the Type II error probability.

COMPLETE SOLUTIONS

Exercise 15.3

a) Errors due to undercoverage (as is the case here) are *not* included in the margin of error.

b) Errors due to nonresponse (as is the case here) are *not* included in the margin of error.

c) The margin of error of a confidence interval only accounts for the chance variation due to random sampling. So the error here *is* included in the margin of error.

Exercise 15.6

See the guided solutions for a full explanation of the way sample size can change your *P*-values.

a) The *z*-test statistic is

$$z = \frac{\bar{x} - \mu_0}{\sigma / \sqrt{n}} = \frac{478 - 475}{100 / \sqrt{100}} = 0.3$$

and

$$P\text{-value} = P(Z > 0.3) = 1 - 0.6179 = 0.3821$$

because the alternative is one-sided.

b) The test statistic is

$$z = \frac{\bar{x} - \mu_0}{\sigma / \sqrt{n}} = \frac{478 - 475}{100 / \sqrt{1000}} = 0.95$$

and

$$P\text{-value} = P(Z > 0.95) = 1 - 0.8289 = 0.1711$$

c) The test statistic is

$$z = \frac{\bar{x} - \mu_0}{\sigma / \sqrt{n}} = \frac{478 - 475}{100 / \sqrt{10000}} = 3$$

and

$$P\text{-value} = P(Z > 3) = 1 - 0.9987 = 0.0013$$

Exercise 15.9

a) A *P*-value of 0.01 means that the probability a subject would do so well when merely guessing is only 0.01. Among 500 subjects, all of whom are merely guessing, we would therefore expect 1%, or 5, of them to do significantly better than random guessing ($P < 0.01$). Thus in 500 tests it is not unusual to see four results with *P*-values on the order of 0.01, even if all are guessing and none have ESP.

b) These four subjects only should be retested with a new, well-designed test. If all four again have low *P*-values (say, below 0.01 or 0.05), we have real evidence that they are not merely guessing. In fact, if any one of the subjects has a very low *P*-value (say, below 0.01), it would also be reasonably compelling evidence that the individual is not merely guessing. A single *P*-value on the order of 0.10, however, would not be particularly convincing.

Exercise 15.11

We begin by rewriting the rule for rejecting H_0 in terms of $\bar{x}$. The rule is to reject H_0 if $z \le -1.645$, or

$$\frac{\bar{x} - 300}{3 / \sqrt{6}} = \frac{\bar{x} - 300}{1.22} \le -1.645.$$

Rewriting this in terms of $\bar{x}$ gives the following rule for rejection. Reject H_0 if

$$\bar{x} \le (1.22)(-1.645) + 300 = 297.99$$

a) If $\mu = 299$, the sampling distribution of $\bar{x}$ is normal with mean 299 and standard deviation $\frac{3}{\sqrt{6}} = 1.22$. The power is the probability of rejecting H_0 when the particular alternative $\mu = 299$ is true. This probability is

$$\text{Power} = P(\bar{x} \leq 297.99) = P\left(\frac{\bar{x}-299}{1.22} \leq \frac{297.99-299}{1.22}\right) = P(Z \leq -0.83) = .2033$$

b) If $\mu = 295$, the sampling distribution of $\bar{x}$ is normal with mean 295 and standard deviation $\frac{3}{\sqrt{6}}$ = 1.22. The power is the probability of rejecting H_0 when the particular alternative $\mu = 295$ is true. This probability is

$$\text{Power} = P(\bar{x} \leq 297.99) = P\left(\frac{\bar{x}-295}{1.22} \leq \frac{297.99-295}{1.22}\right) = P(Z \leq 2.45) = .9929$$

c) The power will be greater than in part (b). We notice in parts (a) and (b) that as μ decreased, the power increased. Thus we would suspect that the power will be greater when μ decreases further to 290. A more precise argument is the following. If $\mu = 290$, $\bar{x}$ is likely to be close to 290 (within a standard deviation or two of 290). 297.99 is several multiples of the standard deviation of 1.22 above 290. Thus it is almost certain that $\bar{x}$ will be less than 297.99 and more likely to be less than 297.99 than if μ was 295 (which is closer to 297.99 than 290 is). This means the power will be very close to 1 and greater than in part (b).

Exercise 15.13

a) The two hypotheses are

H_0: the patient has no medical problem
H_a: the patient has a medical problem

One possible error is to decide

H_a: the patient has a medical problem

when, in fact, the patient does not really have a medical problem. This is a Type I error and in this setting could be called a false positive. The other type of error is to decide

H_0: the patient has no medical problem

when, in fact, the patient does have a problem. This is a Type II error and in this setting could be called a false negative.

b) Most likely we would choose to decrease the error probability for a Type II error, or the false negative probability. Failure to detect a problem (particularly a major problem) when one is present could result in serious consequences (such as death). While a false positive can also have serious consequences (painful or expensive treatment that is not necessary), it is not likely to lead to the kinds of consequences that a false negative could produce. For example, consider the consequences of failure to detect a heart attack, the presence of AIDS, or the presence of cancer. Note, there are cases where some might argue that a false positive would be a more serious error than a false negative. For example, a false positive in a test for Down's Syndrome or a birth defect in an unborn baby might lead parents to consider an abortion. Some would consider this a much more serious error than to give birth to a child with a birth defect.

Exercise 15.15

a) If $\mu = 0$, the sampling distribution of $\bar{x}$ is normal with mean $\mu = 0$ and standard deviation $\frac{\sigma}{\sqrt{n}} = \frac{1}{\sqrt{9}} =$ 0.33. Thus the probability of a Type I error is the probability that $\bar{x} > 0$ when the null hypothesis is true. Computing the z score for $\bar{x}$ we get

$$P(\bar{x} > 0) = P\left(\frac{\bar{x} - 0}{0.33} > \frac{0 - 0}{0.33}\right) = P(Z > 0) = 0.5$$

b) If $\mu = 0.3$, the sampling distribution of $\bar{x}$ is normal with mean $\mu = 0.3$ and standard deviation $\frac{\sigma}{\sqrt{n}} =$ $\frac{1}{\sqrt{9}} = 0.33$. We accept H_0 if $\bar{x} \leq 0$. Thus the probability of a Type II error when $\mu = 0.3$ is

$$P(\bar{x} \leq 0) = P\left(\frac{\bar{x} - 0.3}{0.33} \leq \frac{0 - 0.3}{0.33}\right) = P(Z \leq -0.91) = .1814$$

c) If $\mu = 1$, the sampling distribution of $\bar{x}$ is normal with mean $\mu = 1$ and standard deviation $\frac{\sigma}{\sqrt{n}} = \frac{1}{\sqrt{9}} =$ 0.33. We accept H_0 if $\bar{x} \leq 0$. Thus the probability of a Type II error when $\mu = 1$ is

$$P(\bar{x} \leq 0) = P\left(\frac{\bar{x} - 1}{0.33} \leq \frac{0 - 1}{0.33}\right) = P(Z \leq -3.0) = .0013$$

Exercise 15.33

Recall that the test in Exercise 15.11 used a significance level of 5% to test the hypotheses

H_0: $\mu = 300$
H_a: $\mu < 300$

The probability of a Type I error is the same as the significance level and hence is 0.05.

The probability of a Type II error at the alternative $\mu = 295$ is the probability of accepting the null hypothesis H_0: $\mu = 300$. One minus this probability is the probability of (correctly) rejecting the null hypothesis H_0: $\mu = 300$ when the alternative $\mu = 295$ is true. This last probability is the power at the alternative $\mu = 295$. We found this to be .9929 in part (b) of Exercise 15.11 of this Study Guide. One minus this value is thus the Type II error. Hence the Type II error is $1 - .9929 = .0071$.

CHAPTER 16

INFERENCES ABOUT A POPULATION MEAN

OVERVIEW

Confidence intervals and significance tests for the mean μ of a normal population are based on the sample mean $\bar{x}$ of an SRS. When the sample size n is large, the central limit theorem suggests that these procedures are approximately correct for other population distributions. In Chapters 13 and 14 of your text, the (unrealistic) situation was considered in which we knew the population standard deviation, σ. In this chapter, we consider the more realistic case where σ is not known and we must estimate σ from our SRS by the sample standard deviation s. In Chapters 13 and 14 we used the **one-sample z statistic**

$$z = \frac{\bar{x} - \mu}{\sigma/\sqrt{n}}$$

which has the $N(0,1)$ distribution. Replacing σ by s, we now use the **one-sample t statistic**

$$t = \frac{\bar{x} - \mu}{s/\sqrt{n}}$$

which has the **t distribution** with $n - 1$ **degrees of freedom**. For every positive value of k there is a t distribution with k degrees of freedom, denoted $t(k)$. All are symmetric, bell-shaped distributions, similar in shape to normal distributions but with greater spread. As k increases, $t(k)$ approaches the $N(0,1)$ distribution.

A level C **confidence interval for the mean** μ of a normal population when σ is unknown is

$$\bar{x} \pm t^* \frac{s}{\sqrt{n}}$$

where t^* is the upper $(1 - C)/2$ critical value of the $t(n - 1)$ distribution whose value can be found in Table C in the Appendix of your text or from statistical software. The one-sample t confidence interval has the form estimate $\pm t^* \text{SE}_{\text{estimate}}$, where "SE" stands for **standard error**.

Significance tests of H_0: $\mu = \mu_0$ are based on the one-sample t statistic. P-values or fixed significance levels are computed from the $t(n - 1)$ distribution using Table C or, more commonly in practice, using statistical software.

One application of these one-sample t procedures is to the analysis of data from **matched pairs** studies. We compute the differences between the two values of a matched pair (often before and after measurements on the same unit) to produce a single sample value. The sample mean and standard deviation of these differences are computed. Depending on whether we are interested in a confidence interval or a test of significance concerning the difference in the population means of matched pairs, we either use the one-sample confidence interval or the one-sample significance test based on the t statistic.

For larger sample sizes, the t procedures are fairly **robust** against nonnormal populations. As a rule of thumb, t procedures are useful for nonnormal data when $n \geq 15$ unless the data show outliers or strong skewness, and for samples of size $n \geq 40$, t procedures can be used for even clearly skewed distributions. For smaller samples, it is a good idea to examine stemplots or histograms before you use the t-procedures to check for outliers or skewness.

GUIDED SOLUTIONS

Exercise 16.7

KEY CONCEPTS - one-sample t confidence intervals, checking assumptions

a) With a sample size of only $n = 9$, the most sensible graph would be a stemplot. Complete the stemplot below. Use split stems and just use the numbers to the left of the decimal place.

4 |
5 |
5 |
6 |
6 |

b) To compute a level C confidence interval we use the formula $\bar{x} \pm t^* \frac{s}{\sqrt{n}}$ where t^* is the upper $(1 - C)/2$ critical value of the $t(n - 1)$ distribution, which can be found in Table C. Fill in the missing values below. Don't foget to subtract one from the sample size when finding the appropriate degrees of freedom for the t confidence interval.

$C =$
$n =$
$t^* =$

Now compute the values of $\bar{x}$ and s from the data given. Use statistical software or a calculator.

$\bar{x} =$ $s =$

Substitute all these values into the formula to complete the computation of the 95% confidence interval.

$$\bar{x} \pm t^* \frac{s}{\sqrt{n}} =$$

Exercise 16.11

KEY CONCEPTS - matched pairs experiments, one-sample t tests

a) This is a matched pairs experiment. The matched pair of observations are the right and left hand times on each subject. To avoid confounding with time of day, we would probably want subjects to use both knobs in the same session. We would also want to randomize which knob the subject uses first. How might you do this randomization? What about the order in which the subjects are tested?

b) The project hopes to show that right-handed people find right-hand threads easier to use than left-hand threads. In terms of the mean μ for the population of differences

(left thread time) - (right thread time)

what do we wish to show? This would be the alternative. What are H_0 and H_a (in terms of μ)?

H_0: H_a:

c) For data from a matched pairs study, we compute the differences between the two values of a matched pair to produce a single sample value. These differences are given below for our data.

Right thread	Left thread	Difference = Left - Right
113	137	24
105	105	0
130	133	3
101	108	7
138	115	-23
118	170	52
87	103	16
116	145	29
75	78	3
96	107	11
122	84	-38
103	148	45
116	147	31
107	87	-20
118	166	48
103	146	43
111	123	12
104	135	31
111	112	1
89	93	4
78	76	-2
100	116	16
89	78	-11
85	101	16
88	123	35

The sample mean and standard deviation of these differences need to be computed. Fill in their values in the space provided. Use statistical software or a calculator.

$\bar{x}$ = s =

We now use the one sample significance test based on the t statistic. What value of μ_0 should be used?

$$t = \frac{\bar{x} - \mu_0}{s/\sqrt{n}} =$$

From the value of the t statistic and Table C (or using statistical software), the P-value can be computed. If using Table C, between what two values does the P-value lie?

$\leq$ P-value $\leq$

Exact P-value from software =

What conclusion do you draw?

Note: This problem is most easily done directly using statistical software. The software will compute the differences, the t statistic and the P-value for you. Consult your users manual to see how to do one-sample t tests.

Exercise 16.13

KEY CONCEPTS - matched pairs experiments, confidence intervals

Taking the 25 differences (left – right), we get the mean and standard deviation of the differences as $\bar{x}$ = 13.32, s = 22.94 (See Exercise 16.11 of this study guide). To compute a level C confidence interval we use the formula

$$\bar{x} \pm t^* \frac{s}{\sqrt{n}} =$$

where t^* is the upper $(1 - C)/2$ critical value of the $t(n - 1)$ distribution, which can be found in Table C. Substitute all these values into the formula above to complete the computation of the 95% confidence interval. Don't forget to subtract one from the sample size when finding the appropriate degrees of freedom for the t confidence interval.

As an alternative to computing the mean of the differences, you could evaluate the ratio of the mean time for right-hand threads as a percent of left-hand threads to help determine if the time saved is of practical importance.

$$\bar{x}_R \,/\, \bar{x}_L =$$

Exercise 16.21

KEY CONCEPTS - one-sample t confidence intervals, checking assumptions

a) Recall some of the graphical methods of Chapter 1 of your text for describing data. We have drawn a boxplot below. Complete the histogram on the right using the class intervals provided.

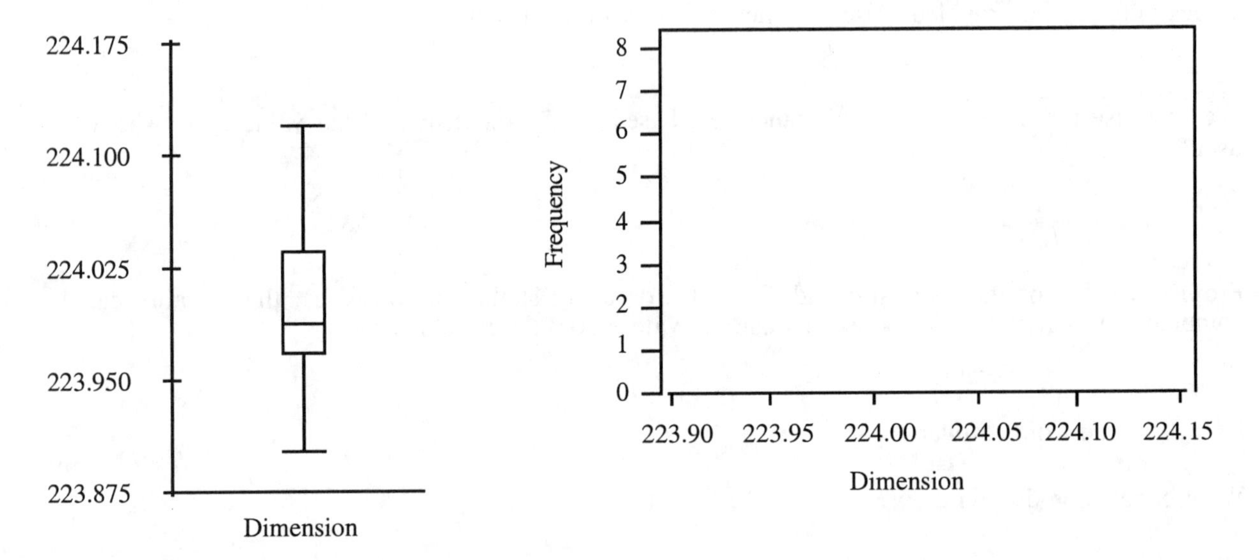

Using the histogram and/or the boxplot, is there evidence for outliers or strong skewness? Given the general guidelines on the robustness of the t procedures, do you think it is valid to use these procedures for these data?

b) We are interested in whether the data provide evidence that the mean dimension is not 224 mm. State the appropriate null and alternative hypotheses below.

H_0:
H_a:

From the data, first calculate the sample mean and the sample standard deviation, preferably using statistical software or a calculator. Fill in the values below.

$\bar{x}$ = s =

Now use these to compute the value of the one-sample t statistic.

$$t = \frac{\bar{x} - \mu_0}{s/\sqrt{n}} =$$

Compute the P-value using Table C or software. How many degrees of freedom are there? The degrees of freedom tell you which row of Table C you need to refer to for critical values. Remember, if the alternative is two-sided then the probability found in the table needs to be doubled.

degrees of freedom:

P-value:

Conclusion:

Exercise 16.32

KEY CONCEPTS - confidence intervals based on the one-sample t statistic, assumptions underlying t procedures

a) To compute a level C confidence interval we use the formula $\bar{x} \pm t^* \frac{s}{\sqrt{n}}$ where t^* is the upper $(1 - C)/2$ critical value of the $t(n - 1)$ distribution, which can be found in Table C. Fill in the missing values below. Don't foget to subtract one from the sample size when finding the appropriate degrees of freedom for the t confidence interval.

C =
n =
t^* =

The values of $\bar{x}$ and s are given in the problem.

$\bar{x}$ = s =

Substitute all these values into the formula to complete the computation of the 95% confidence interval.

$$\bar{x} \pm t^* \frac{s}{\sqrt{n}} =$$

b) What are the assumptions required for the t confidence interval. Which assumptions are satisfied and which may not be? How were the subjects in the study obtained? How were the subjects in the placebo group obtained?

COMPLETE SOLUTIONS

Exercise 16.7

a) The stemplot is given below. There are no outliers and the plot is skewed left. With this few observations, it is difficult to check the assumptions. In this case we might still use the t procedures, but not with as much confidence in their validity as we had in other examples.

```
4 | 9
5 | 1 4
5 |
6 | 0 3 3 4 4
6 | 5
```

b) An approximate 95% confidence interval for the mean percent of nitrogen in ancient air can be calculated from the data on the 9 specimens of amber. We use the formula for a t interval, namely $\bar{x} \pm t^* \frac{s}{\sqrt{n}}$. In this problem, $\bar{x} = 59.589$, $s = 6.2553$, $n = 9$, hence t^* is the upper $(1 - 0.95)/2 = 0.025$ critical value for the $t(8)$ distribution. From Table C we see $t^* = 2.306$. Thus the 95% confidence interval is

$$59.589 \pm 2.306 \frac{6.2553}{\sqrt{9}} = 59.589 \pm 4.808 = (54.78,\ 64.40)$$

Many statistical software packages will compute a confidence interval for you directly, after you input the data.

Exercise 16.11

a) The randomization might be carried out by simply flipping a fair coin. If the coin comes up heads, use the right-hand threaded knob first. If the coin comes up tails, use the left-hand threaded knob first. Alternatively, in order to balance out the number of times each type is used first, one might choose an SRS of 12 of the 25 subjects. These 12 use the right-hand thread knob first. Everyone else uses the left-hand thread knob first.

A second place one might use randomization is in the order in which subjects are tested. Use a table of random digits to determine this order. Label subjects 01 to 25. The first label that appears in the list of random digits (read in groups of two digits) is the first subject measured. The second label which appears, the next subject measured, etc. This randomization is probably less important than the one described in the previous paragraph. It would be important if the order or time at which a subject was tested might have an effect on the measured response. For example, if the study began early in the morning, the first subject might be sluggish if still sleepy. Sluggishness might lead to longer times and perhaps a larger difference in times. Subjects tested later in the day might be more alert.

b) In terms of μ, the mean of the population of differences, (left thread time) – (right thread time), we wish to test if the times for the left threaded knobs are longer than for the right threaded knobs, i.e.

H_0: $\mu = 0$ and H_a: $\mu > 0$

c) For the 25 differences we compute

$\bar{x} = 13.32$ $\qquad s = 22.94.$

We then use the one sample significance test based on the t statistic.

$$t = \frac{\bar{x} - \mu_0}{s/\sqrt{n}} = \frac{13.32 - 0}{22.94/\sqrt{25}} = 2.903.$$

From the value of the t statistic and Table C the P-value is between 0.0025 and 0.005.

df = 24

p	.005	.0025
t^*	2.797	3.091

Using statistical software the P-value is computed as P-value = 0.0039.

We conclude that there is strong evidence that the time for left-hand threads is greater than for right-hand threads on average.

Exercise 16.13

$\bar{x} = 13.32$, $s = 22.94$, $n = 25$, and t^* is the upper $(1 - 0.90)/2 = 0.05$ critical value for the $t(24)$ distribution. From Table D, we see that $t^* = 1.711$. Thus, the 90% confidence interval is

$$13.32 \pm 1.711 \frac{22.94}{\sqrt{25}} = 13.32 \pm 7.85 = (5.47, 21.17)$$

Computing the means, $\bar{x}_R = 104.12$, $\bar{x}_L = 117.44$, and $\bar{x}_R / \bar{x}_L = 88.7\%$, so those using the right-handed threads complete the task in about 90% of the time it takes those using the left-handed threads. (As an alternative, if for each subject we first take the ratio [right-thread] / [left-thread] and then average these ratios, we get 91.7%, which is almost the same answer.)

Exercise 16.21

a) The boxplot was provided in the guided solutions and the completed histogram is given below (one could also make a stemplot of the data).

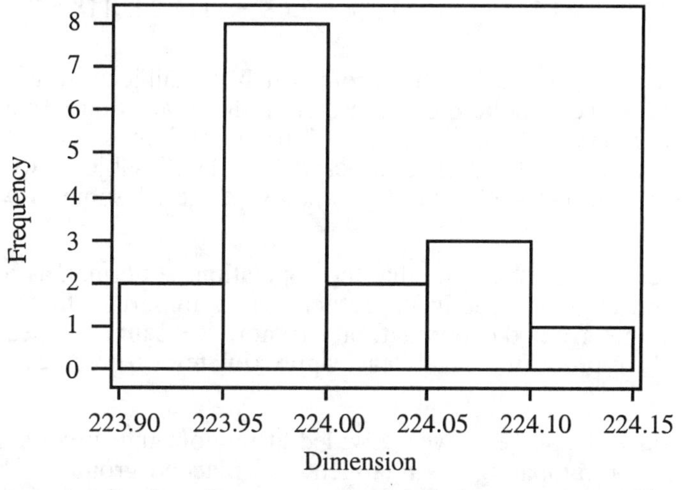

Using either the histogram or the boxplot, we can see that there are no outliers in the data. The data appear a bit skewed to the right, but not so strongly as to threaten the validity of the t procedure given that the sample size is 16 (in the section on the robustness of t procedures, t procedures are safe for samples of size $n \geq 15$ unless there are outliers and/or strong skewness).

b) Since we are interested in whether the data provide evidence that the mean dimension is not 224 mm. (no direction of the difference is specified), we wish to test the hypotheses

$$H_0: \mu = 224 \text{ mm.}$$
$$H_a: \mu \neq 224 \text{ mm.}$$

From the data we calculate the basic statistics to be

$\bar{x} = 224.0019$, $s = 0.618$ and the standard error as $\dfrac{s}{\sqrt{n}} = \dfrac{0.0618}{\sqrt{16}} = 0.01545.$

Substituting these in the formula for t yields

$$t = \frac{\bar{x} - \mu_0}{s/\sqrt{n}} = \frac{224.0019 - 224}{0.01545} = 0.123$$

The P-value for $t = 0.123$ is twice the area to the right of 0.123 under the t distribution curve with $n - 1 = 15$ degrees of freedom. Using Table C, we search the df = 15 row for entries that bracket 0.123. Since 0.123 lies to the left of the smallest entry in the Table corresponding to a probability of 0.25, the P-value is

df = 15

p	.25
t^*	0.691

therefore greater than .25 × 2 = 0.50 for this two-sided test. The data do not provide strong evidence that the mean differs from 224 mm. (Computer software gives a P-value of 0.9038).

Exercise 16.32

a) A 95% confidence interval for the mean systolic blood pressure in the population from which the subjects were recruited can be calculated from the data on the 27 members of the placebo group, since these are randomly selected from the 54 subjects. We use the formula for a t interval, namely $\bar{x} \pm t^* \frac{s}{\sqrt{n}}$. In this problem, $\bar{x} = 114.9$, $s = 9.3$, $n = 27$, hence t^* is the upper $(1 - 0.95)/2 = 0.025$ critical value for the $t(26)$ distribution. From Table C we see $t^* = 2.056$. Thus the 95% confidence interval is

$$114.9 \pm 2.056 \frac{9.3}{\sqrt{27}} = 114.9 \pm 3.68 = (111.22, 118.58)$$

b) For the procedure used in (a), the population from which the subjects were drawn should be such that the distribution of the seated systolic blood pressure in the population is normal. The 27 subjects used for the confidence interval in part (a) should be a random sample from this population. Unfortunately, we do not know if the latter is the case. While 27 subjects were selected at random from the total of 54 subjects in the study, we do not know if the 54 subjects were a random sample from this population.

With a sample of 27 subjects, it is not crucial that the population be normal, as long as the distribution is not strongly skewed and the data contain no outliers. It is important that the 27 subjects can be considered a random sample from the population. If not, we cannot appeal to the central limit theorem to insure that the t procedure is at least approximately correct even if the data are not normal.

(Note: It turns out that since the subjects were divided at random into treatment and control groups, there do exist procedures for comparing the treatment and placebo groups. These are not based on the t distribution, but are valid as long as treatment groups are determined by randomization. However, the conclusions drawn from these procedures apply only to the subjects in the study. To generalize the conclusions to a larger population, we must know that the subjects are a random sample from this larger population.)

CHAPTER 17

TWO-SAMPLE PROBLEMS

OVERVIEW

One of the the most commonly used significance tests is the **comparison of two population means,** μ_1 and μ_2. In this setting we have two distinct, independent SRS from two populations or two treatments in a randomized comparative experiment. The procedures are based on the difference $\bar{x}_1 - \bar{x}_2$. When the populations are not normal, the results obtained using the methods of this section are approximately correct due to the central limit theorem.

Tests and confidence intervals for the difference in the population means, $\mu_1 - \mu_2$, are based on the **two-sample *t* statistic**. Despite the name, this test statistic does *not* have an exact *t* distribution. However there are good approximations to its distribution which allow us to carry out valid significance tests. Conservative procedures use the $t(k)$ distribution as an approximation where the degrees of freedom k is taken to be the smaller of $n_1 - 1$ and $n_2 - 1$. More accurate procedures use the data to estimate the degrees of freedom k. This is the procedure that is followed by most statistical software.

To carry out a significance test for $H_0\colon \mu_1 = \mu_2$, use the two-sample t statistic

$$t = \frac{(\bar{x}_1 - \bar{x}_2)}{\sqrt{\dfrac{s_1^2}{n_1} + \dfrac{s_2^2}{n_2}}}$$

The P-value is found by using the approximate distribution $t(k)$, where k is estimated from the data when using statistical software, or can be taken to be the smaller of $n_1 - 1$ and $n_2 - 1$ for a conservative procedure. An approximate confidence C level **confidence interval** for $\mu_1 - \mu_2$ is given by

$$(\bar{x}_1 - \bar{x}_2) \pm t^* \sqrt{\frac{s_1^2}{n_1} + \frac{s_2^2}{n_2}}$$

where t^* is the upper $(1 - C)/2$ critical value for $t(k)$, where k is estimated from the data when using statistical software, or can be taken to be the smaller of $n_1 - 1$ and $n_2 - 1$ for a conservative procedure. The procedures are most robust to failures in the assumptions when the sample sizes are equal.

The **pooled two-sample *t* procedures** are used when we can safely assume that the two populations have equal variances. The modifications in the procedure are the use of the pooled estimator of the common unknown variance and critical values obtained from the $t\,(n_1 + n_2 - 2)$ distribution.

There are formal inference procedures to compare the standard deviations of two normal populations as well as the two means. The validity of the procedures is seriously affected when the distributions are nonnormal, and they are not recommended for regular use. The procedures are based on the ***F* statistic,** which is the ratio of the two sample variances

$$F = \frac{s_1^2}{s_2^2}.$$

If the data consist of independent simple random samples of sizes n_1 and n_2 from two normal populations, then the F statistic has the F distribution, $F(n_1 - 1, n_2 - 1)$, if the two population standard deviations σ_1 and σ_2 are equal. Critical values of the F distribution are provided in Table D. Because of the skewness of the F distribution, when carrying out the two-sided test we take the ratio of the larger to the smaller standard deviation, which eliminates the need for lower critical values.

GUIDED SOLUTIONS

Exercise 17.3

KEY CONCEPTS - single sample, matched pairs or two-samples

Are there one or two samples involved? Was matching done?

Exercise 17.4

KEY CONCEPTS - single sample, matched pairs or two-samples

Are there one or two samples involved? Was matching done?

Exercise 17.5

KEY CONCEPTS - single sample, matched pairs or two-samples

a) Recall that $\text{SEM} = \frac{s}{\sqrt{n}}$ is the standard error of the mean. The description of the study provides the sample sizes. The means are given and you should be able to reproduce the standard deviation s from the value of SEM and the sample size for each group. Fill in your answers in the table below.

Group	Treatment	n	$\bar{x}$	s
1	IDX			
2	Untreated			

b) Based on the two sample sizes, what value would you use for the degrees of freedom in the conservative two-sample t procedure?

Exercise 17.13

KEY CONCEPTS - tests using the two-sample t, t approximation

The summary statistics obtained from the TI-83 output for the comparison of the measured variable (height of the second spike as a percent of the first) between poisoned and unpoisoned rats are reproduced on the next page.

Group	Treatment	n	$\bar{x}$	s
1	Poisoned	6	17.6000	6.3401
2	Unpoisoned	6	9.4998	1.9501

The values of the t statistic and the approximate degrees of freedom are provided in the output of the TI-83. These summary statistics are all you need to verify the calculations.

$$t = \frac{\bar{x}_1 - \bar{x}_2}{\sqrt{\frac{s_1^2}{n_1} + \frac{s_2^2}{n_2}}} =$$

$$df = \frac{\left(\frac{s_1^2}{n_1} + \frac{s_2^2}{n_2}\right)^2}{\frac{1}{n_1 - 1}\left(\frac{s_1^2}{n_1}\right)^2 + \frac{1}{n_2 - 1}\left(\frac{s_2^2}{n_2}\right)^2} =$$

Exercise 17.15

KEY CONCEPTS - tests using the two-sample t, interpreting results

All of the calculations have been performed by the TI-83 and are provided in the output. Using this information, write a summary in a sentence or two. Be sure to include the t, df, P and a conclusion.

Exercise 17.19

KEY CONCEPTS - the F test for equality of the standard deviations of two normal populations

The data are from Exercise 7.5 of your text, and are measurements from a randomized comparative experiment to study the effect of IDX versus a control in the treatment of scrapie in hamsters. The summary statistics for the comparison of the lifetimes of the two groups are reproduced below.

Group	Treatment	n	$\bar{x}$	s
1	IDX	10	116	17.71
2	Untreated	10	88.5	6.01

We are interested in testing the hypotheses $H_0: \sigma_1 = \sigma_2$ and $H_a: \sigma_1 \neq \sigma_2$. The two-sided test statistic is the larger variance divided by the smaller variance and if σ_1 and σ_2 are equal has the $F(n_1 - 1, n_2 - 1)$. Remember that n_1 is the numerator sample size. To organize your calculations, first compute the three quantities below.

$$F = \frac{\text{larger } s^2}{\text{smaller } s^2} =$$

Numerator df $(n_1 - 1)$ =

Denominator df $(n_2 - 1)$ =

Compare the value of the F you computed to the critical values given in Table D, making sure to go to the row and column corresponding to the appropriate degrees of freedom, or as close as you can get to these degrees of freedom if they are not in the table. Between which two critical values does F lie? What can be said about the P-value from the table? If available, use software to calculate the exact P-value.

Exercise 17.29

KEY CONCEPTS - tests using the two-sample t, checking assumptions, back-to-back stemplots

a) For two small data sets, the simplest graphical display is the back-to-back stemplot (see, for example, Exercise 1.26 of this study guide). It allows a simple comparison of the two data sets and will show outliers skewness. Complete the back-to-back stemplot. We have filled in the three smallest men's and the three smallest women's scores. Are there outliers and/or skewness? Are t procedures still appropriate? Explain.

Women		Men
	7	05
	8	8
	9	
931	10	
	11	
	12	
	13	
	14	
	15	
	16	
	17	
	18	
	19	
	20	

b) Letting μ_W and μ_M represent the population means for women and men respectively, first state the hypotheses of interest in terms of these parameters.

H_0: H_a:

Below are the summary statistics. You can use a calculator or statistical software to verify these values. Now use these summary statistics to compute the value of t below.

Group	n	$\bar{x}$	s
Women	18	141.056	26.4363
Men	20	121.250	32.8519

$$t = \frac{\bar{x}_W - \bar{x}_M}{\sqrt{\frac{s_W^2}{n_W} + \frac{s_M^2}{n_M}}} =$$

Using the conservative degrees of freedom, what can you say about the P-value? What do you conclude about the SSHA scores of men versus women?

Note that if you did this problem by entering the data into statistical software, the value of the t statistic should be the same, but the degrees of freedom used will be different and the P-value will differ slightly.

Exercise 17.31

KEY CONCEPTS - two-sample t confidence interval

The formula for the 90% confidence interval is $\bar{x}_1 - \bar{x}_2 \pm t^* \sqrt{\frac{s_1^2}{n_1} + \frac{s_2^2}{n_2}}$, where t^* is the upper $(1 - C)/2 = 0.05$ critical value for the t distribution with degrees of freedom equal to the smaller of $n_W - 1$ and $n_M - 1$. The means and standard deviations are given in the guided solution to Exercise 17.29 in this study guide. Don't forget to square the standard deviations in the formula for the standard error. Complete the calculations in steps as suggested below.

$t^* =$

$\sqrt{\frac{s_1^2}{n_1} + \frac{s_2^2}{n_2}} =$

$\bar{x}_1 - \bar{x}_2 \pm t^* \sqrt{\frac{s_1^2}{n_1} + \frac{s_2^2}{n_2}} =$

COMPLETE SOLUTIONS

Exercise 17.3

This example involves a single sample. We have a sample of 20 measurements and we want to see if the mean for this method agrees with the known concentration.

Exercise 17.4

This example involves two samples, the set of measurements on each method. Note that we are not told of any matching.

Exercise 17.5

a) There are 20 infected hamsters with 10 assigned at random to each group. The sample sizes are then 10 for the IDX group and 10 for the untreated group. The means are given and since $\text{SEM} = \frac{s}{\sqrt{n}}$, the standard deviation for each group is $s = (\text{SEM})\sqrt{n}$. For the IDX group this gives $s = (\text{SEM})\sqrt{n} = 5.6\sqrt{10} = 17.71$ and for the untreated group $s = (\text{SEM})\sqrt{n} = 1.9\sqrt{10} = 6.01$.

Group	Treatment	n	$\bar{x}$	s
1	IDX	10	116	17.71
2	Untreated	10	88.5	6.01

b) The degrees of freedom are the smaller of $n_1 - 1 = 10 - 1 = 9$ and $n_2 - 1 = 10 - 1 = 9$ which is 9.

Exercise 17.13

Entering the summary statistics into the formulas for the t statistic and the approximate degrees of freedom gives

$$t = \frac{\bar{x}_1 - \bar{x}_2}{\sqrt{\frac{s_1^2}{n_1} + \frac{s_2^2}{n_2}}} = \frac{17.6000 - 9.4998}{\sqrt{\frac{(6.3401)^2}{6} + \frac{(1.9501)^2}{6}}} = 2.99$$

and

$$df = \frac{\left(\frac{s_1^2}{n_1} + \frac{s_2^2}{n_2}\right)^2}{\frac{1}{n_1 - 1}\left(\frac{s_1^2}{n_1}\right)^2 + \frac{1}{n_2 - 1}\left(\frac{s_2^2}{n_2}\right)^2} = \frac{\left(\frac{6.3401^2}{6} + \frac{1.9501^2}{6}\right)^2}{\frac{1}{6-1}\left(\frac{6.3401^2}{6}\right)^2 + \frac{1}{6-1}\left(\frac{1.9501^2}{6}\right)^2} = 5.9$$

Exercise 17.15

Using the information in the TI-83 output, we have $t = 2.9912$, df = 5.9 and the P-value = 0.0246. We would reject the null hypothesis at a significance level of 0.05, since the P-value falls below 0.05, but not at the 0.01 significance level. We conclude that there is good evidence of a difference in the mean of the measured variable (height of the second spike as a percent of the first) between poisoned and unpoisoned rats, with the poisoned rats tending to have larger values of the measured variable.

Exercise 17.19

Since the treated group has the larger standard deviation (and variance), this is the variance that goes in the numerator, so that the degrees of freedom are $n_1 - 1 = 10 - 1 = 9$, and $n_2 - 1 = 10 - 1 = 9$. To compute the value of the test statistic, note that it is the standard deviations and not the variances that are given as summary statistics. The standard deviations must first be squared to get the two variances. The larger standard deviation is for the IDX group and the sample variance for this group is $(17.71)^2 = 313.6441$. For the untreated group, the sample variance is $(6.01)^2 = 36.1201$. The value of the test statistic is

$$F = \frac{\text{larger } s^2}{\text{smaller } s^2} = \frac{313.6441}{36.1201} = 8.68.$$

Since the degrees of freedom (9, 9) are in the table, we go to Table D and find the two values that bracket the computed value of $F = 8.68$.

df = (9, 9)		
p	0.01	0.001
F^*	5.26	9.89

Since the test is two-sided, we double the significance levels from the table and conclude the P-value is between 0.002 and 0.02. Using computer software we find the P-value to be 0.0036. This is significant evidence of unequal standard deviations between the two groups.

Exercise 17.29

a) Both of the stemplots show distributions that are slightly skewed to the right and each has one or two moderate high outliers. Since the sum of the sample sizes is close to 40 a t procedure may be used, but with some caution.

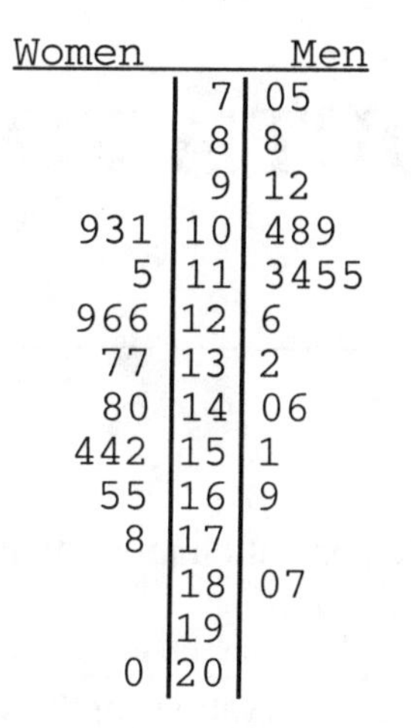

Women Men
7 | 05
8 | 8
9 | 12
931 | 10 | 489
5 | 11 | 3455
966 | 12 | 6
77 | 13 | 2
80 | 14 | 06
442 | 15 | 1
55 | 16 | 9
8 | 17 |
18 | 07
19 |
0 | 20 |

b) We test $H_0: \mu_W = \mu_M$ versus $H_a: \mu_W > \mu_M$. From the summary statistics the value of t is computed as

$$t = \frac{\bar{x}_W - \bar{x}_M}{\sqrt{\frac{s_W^2}{n_W} + \frac{s_M^2}{n_M}}} = \frac{141.056 - 121.250}{\sqrt{\frac{(26.4363)^2}{18} + \frac{(32.8519)^2}{20}}} = 2.06$$

Since the degrees of freedom are the smaller of $n_W - 1 = 17$ and $n_M - 1 = 19$, we go to Table C using df = 17 and find the two values that bracket the computed value of $t = 2.06$.

df = 17

p	0.025	0.05
F^*	2.110	1.740

Because the test is one-sided, $0.025 < P\text{-value} < 0.05$. This gives fairly strong evidence (statistically significant at the 5% level) that men have lower SSHA scores than women.

Exercise 17.31

The upper $(1 - C)/2 = 0.05$ critical value for the t distribution with degrees of freedom equal to the smaller of $n_W - 1 = 17$ and $n_M - 1 = 19$ is $t^* = 1.740$. The standard error is

$$\sqrt{\frac{s_1^2}{n_1} + \frac{s_2^2}{n_2}} = \sqrt{\frac{(26.4363)^2}{18} + \frac{(32.8519)^2}{20}} = 9.633$$

and the 90% confidence interval is

$$\bar{x}_1 - \bar{x}_2 \pm t^* \sqrt{\frac{s_1^2}{n_1} + \frac{s_2^2}{n_2}} = (141.0516 - 121.250) \pm (1.740)(9.633) = (3.05,\ 36.57).$$

CHAPTER 18

INFERENCE ABOUT A POPULATION PROPORTION

OVERVIEW

In this chapter we consider inference about a population proportion p based on the **sample proportion**

$$\hat{p} = \frac{\text{count of successes in the sample}}{\text{count of observations in the sample}}$$

obtained from an SRS of size n, where X is the number of "successes" (occurrences of the event of interest) in the sample. To use the methods of this chapter for inference, the following assumptions need to be satisfied.

- The data are an SRS from the population of interest.
- The population is at least 10 times as large as the sample.
- The sample size is sufficiently large. Guidelines for sample sizes are given for each method discussed below.

In this case, we can treat $\hat{p}$ as having a distribution that is approximately normal with mean $\mu = p$ and standard deviation $\sigma = \sqrt{p(1-p)/n}$.

An **approximate level C confidence interval** for p is

$$\hat{p} \pm z^* \sqrt{\frac{\hat{p}(1-\hat{p})}{n}}$$

where z^* is the upper $(1 - C)/2$ critical value of the standard normal distribution,

$$\sqrt{\frac{\hat{p}(1-\hat{p})}{n}}$$

is the **standard error** of $\hat{p}$, and $z^* \sqrt{\frac{\hat{p}(1-\hat{p})}{n}}$ is the **margin of error.** Use this interval only when the counts of successes and failures in the sample are both at least 15.

A more accurate confidence interval for smaller samples is the **plus four confidence interval**. To get this interval, add four imaginary observations, two successes and two failures, to your sample. Then use the formula above for the confidence interval. Use the plus four confidence interval when the confidence level C is at least 90% and the sample size n is at least 10.

The **sample size** n required to obtain a confidence interval of approximate margin of error m for a proportion is

$$n = \left(\frac{z^*}{m}\right)^2 p^*(1-p^*)$$

where p^* is a guessed value for the population proportion and z^* is the upper $(1 - C)/2$ critical value of the standard normal distribution. To guarantee that the margin of error of the confidence interval is less than or equal to m no matter what the value of the population proportion may be, use a guessed value of $p^* = 1/2$, which yields

$$n = \left(\frac{z^*}{2m}\right)^2$$

Tests of the hypothesis H_0: $p = p_0$ are based on the z **statistic**

$$z = \frac{\hat{p} - p_0}{\sqrt{\frac{p_0(1-p_0)}{n}}}$$

with P-values calculated from the $N(0, 1)$ distribution. Use this test when $np_0 \geq 10$ and $n(1 - p_0) \geq 10$.

GUIDED SOLUTIONS

Exercise 18.1

KEY CONCEPTS - parameters and statistics, proportions

a) To what group does the study refer?

Population =

Parameter p =

b) A statistic is a number computed from a sample. What is the size of the sample and how many in the sample said they prayed at least once in a while? From these numbers compute

$\hat{p}$ =

Exercise 18.6

KEY CONCEPTS - when to use the procedures for inference about a proportion

Recall the assumptions needed to safely use the methods of this section to compute a confidence interval:

- The data are an SRS from the population of interest.
- The population is at least 10 times as large as the sample.
- For a confidence interval, n is so large that both the count of successes $n\hat{p}$ and the count of failures $n(1 - \hat{p})$ are 15 or more.

These are the conditions we must check. Are the conditions met?

Exercise 18.11

KEY CONCEPTS - large sample confidence intervals for a proportion

a) Recall the assumptions needed to safely use the methods of this section to compute a confidence interval:

- The data are an SRS from the population of interest.
- The population is at least 10 times as large as the sample.
- For a confidence interval, n is so large that both the count of successes and the count of failures are 15 or more.

These are the conditions we must check. Are the conditions met?

b) An approximate 95% confidence interval for the proportion p of all teens who have a TV set in their room is given by the formula

$$\hat{p} \pm z^* \sqrt{\frac{\hat{p}(1-\hat{p})}{n}}$$

where z^* is the upper 0.025 critical value of the standard normal distribution. What are n, $\hat{p}$, and z^* here?

n = sample size =

$\hat{p}$ = the proportion of respondents who had a TV in their room =

z^* =

Now compute the 95% confidence interval.

$$\hat{p} \pm z^* \sqrt{\frac{\hat{p}(1-\hat{p})}{n}} =$$

c) What is the margin of error stated in the article and how does it compare to the margin of error you found in part (b)?

Exercise 18.15

KEY CONCEPTS - accurate confidence intervals for a proportion

a) We use the plus four confidence interval. To get this interval, first add four imaginary observations, two successes and two failures, to the sample. The number of observations in the sample is 127 and the number of successes is (number that said they prayed at least a few times a year). Thus, we pretend

$n = 127 + 4 =$

number of successes (number that said they prayed at least a few times a year) = 107 + 2 =

Now we use the formula for the confidence interval. An approximate 99% confidence interval for p = the proportion of all students who pray is

$$\hat{p} \pm z^* \sqrt{\frac{\hat{p}(1-\hat{p})}{n}}$$

where z^* is the upper 0.005 critical value of the standard normal distribution. What are n, $\hat{p}$, and z^* here?

n = the "imaginary" sample size from above =

$\hat{p}$ = ("imaginary" number who said they prayed at least a few times a year)/n =

z^* =

Now compute the 99% confidence interval.

$$\hat{p} \pm z^* \sqrt{\frac{\hat{p}(1-\hat{p})}{n}} =$$

b) Is it reasonable to assume that psychology and communication majors are representative of all undergraduates?

Exercise 18.17

KEY CONCEPTS - sample size and margin of error

The sample size n required to obtain a confidence interval of approximate margin of error m for a proportion is

$$n = \left(\frac{z^*}{m}\right)^2 p^*(1-p^*)$$

where p^* is a guessed value for the population proportion and z^* is the critical value of the standard normal distribution for the desired level of confidence. To apply this formula here we must determine

m = desired margin of error =

p^* = a guessed value for the population proportion =

C = desired level of confidence =

z^* = the upper $(1-C)/2$ critical value of the standard normal distribution =

From the statement of the exercise, what are these values? Once you have determined them, use the formula to compute the required sample size n.

$$n = \left(\frac{z^*}{m}\right)^2 p^*(1-p^*) =$$

Exercise 18.19

KEY CONCEPTS - testing hypotheses about a proportion

What statistical hypotheses should you test to answer this question?

You should next check that the appropriate conditions are satisfied, namely,

Is the sample an SRS from a large population?

Is the sample size reasonably large?

Are $np_0 \geq 10$ and $n(1-p_0) \geq 10$?

Now, compute the sample proportion of 13- to 17-year olds with a TV in their room,

$\hat{p}$ =

Now compute

$$z = \frac{\hat{p} - p_0}{\sqrt{\frac{p_0(1-p_0)}{n}}} =$$

and

P-value =

of your test. What do you conclude?

Exercise 18.35

KEY CONCEPTS - testing hypotheses about a proportion, confidence intervals for a proportion

a) First state the hypotheses you will test in terms of p.

Now compute the z-test statistic (identify n and p_0 and calculate $\hat{p}$).

$$z = \frac{\hat{p} - p_0}{\sqrt{\frac{p_0(1 - p_0)}{n}}} =$$

For your hypotheses,

P-value =

Is your result significant at the 5% level? (What must the P-value satisfy for this to be true?)

Finally, state your practical conclusions.

b) Recall that an approximate level C confidence interval for p is

$$\hat{p} \pm z^* \sqrt{\frac{\hat{p}(1 - \hat{p})}{n}}$$

where z^* is the upper $(1 - C)/2$ critical value of the standard normal distribution. From the information given in the problem, provide the following values.

n = sample size =

$\hat{p}$ = sample proportion =

C = level of confidence requested =

z^* = upper $(1-C)/2$ critical value of the standard normal distribution =

Use Table A (or C) to find z^*. Now substitute these values into the formula for the confidence interval to complete the problem.

$$\hat{p} \pm z^* \sqrt{\frac{\hat{p}(1-\hat{p})}{n}} =$$

c) If you are not sure how to answer this, review Chapter 8 of your textbook.

COMPLETE SOLUTIONS

Exercise 18.1

a) The population is presumably all college students. The parameter p is the proportion of all college students who pray at least once in a while.

b) The statistic is $\hat{p}$ the proportion in the sample who said that they prayed at least once in a while. It has value

$$\hat{p} = 107/127 = 0.8425$$

Exercise 18.6

a) The data are an SRS from the population of interest.

The population consists of 175 students. This is *not* at least 10 times as large as the sample size $n = 50$.

The methods of this section *cannot* be safely used.

Exercise 18.11

a) The data are a random sample from the population of interest, all 13- to 17-year olds in the U.S. We can reasonably assume that the total number of 13- to 17-year olds in the U.S. is at least 10 times as large as the sample size of $n = 1048$.

The sample size is $n = 1048$, which is large.

The number of successes (those who had a television in their room) is 692, and the number of failures (those who did not have a television in their room) is 1048 – 692 = 356. Both of these are much bigger than 15.

The appropriate conditions are satisfied.

b) An approximate 95% confidence interval for p = the proportion of all teens (i.e., 13- to 17-year olds) who have a TV set in their room is

$$\hat{p} \pm z^* \sqrt{\frac{\hat{p}(1-\hat{p})}{n}}$$

where z^* is the upper 0.025 critical value of the standard normal distribution, which in this case is 1.96. We have $n = 1048$ and we know that 692 of the respondents had a TV in their room. Thus, the proportion of respondents who had a TV in their room is

$$\hat{p} = 692/1048 = 0.66$$

and our 95% confidence interval is

$$\hat{p} \pm z^* \sqrt{\frac{\hat{p}(1-\hat{p})}{n}} = 0.66 \pm 1.96\sqrt{\frac{0.66(1-0.66)}{1048}} = 0.66 \pm 1.96(0.0146) = 0.66 \pm 0.03$$

or 0.63 to 0.69.

c) The news article quote tells us that the margin of error is 3%. Our margin of error from part (b) is 0.03, which is 3% when converted to a percent. Thus, our results agree with the news article.

Exercise 18.15

a) We use the plus four confidence interval. To get this interval, first add four imaginary observations, two successes and two failures, to the sample. This means that we pretend

$n = 127 + 4 = 131$

number of successes (number that said they prayed at least a few times a year) = 107 + 2 = 109.

Now we use the formula for the confidence interval. An approximate 99% confidence interval for p = the proportion of all students who pray is

$$\hat{p} \pm z^* \sqrt{\frac{\hat{p}(1-\hat{p})}{n}}$$

where z^* is the upper 0.005 critical value of the standard normal distribution, which in this case is $z^* = 2.576$. We pretend $n = 131$ and that 109 of the respondents said they prayed at least a few times a year. Thus, the proportion of respondents who said they prayed at least a few times a year is

$$\hat{p} = 109/131 = 0.83$$

and our 99% confidence interval is

$$\hat{p} \pm z^* \sqrt{\frac{\hat{p}(1-\hat{p})}{n}} = 0.83 \pm 2.576\sqrt{\frac{0.83(1-0.83)}{131}} = 0.83 \pm 2.576(0.0328) = 0.83 \pm 0.08$$

or 0.75 to 0.91.

b) Students choose to be psychology or communication majors for a reason. Their interests and attitudes are likely to differ from those of students who choose other majors. It is probably unreasonable to assume that this is an SRS from the population of all students.

Exercise 18.17

We start with the guess that $p^* = 0.75$. For 95% confidence we use $z^* = 1.96$. The sample size we need for a margin of error $m = 0.04$ is thus

$$n = \left(\frac{z^*}{m}\right)^2 p^*(1-p^*) = \left(\frac{1.96}{0.04}\right)^2 0.75(1-0.75) = 450.1875$$

We round this up to $n = 451$. Thus, a sample of size 451 is needed to estimate the proportion of Americans with at least one Italian grandparent who can taste PTC to within $\pm$ 0.04 with 95% confidence.

Exercise 18.19

We want to determine if more than half of all teens have a TV in their room, so we wish to test the hypotheses

$$H_0: p = 0.5 \qquad H_a: p > 0.5$$

The data are a random sample from the population of interest, all 13- to 17-year olds in the U.S. We can reasonably assume that the total number of 13- to 17-year olds in the U.S. is at least 10 times as large as the sample size of $n = 1048$.

The sample size is $n = 1048$, which is large.

The value of p_0 is 0.5. Thus $np_0 = 1048(0.5) = 524 \geq 10$ and $n(1 - p_0) = 1048(1 - 0.5) = 524 \geq 10$

The appropriate conditions appear to be satisfied.

The sample proportion of 13- to 17-year olds with a TV in their room is

$$\hat{p} = 692/1048 = 0.66$$

and we find

$$z = \frac{\hat{p} - p_0}{\sqrt{\dfrac{p_0(1-p_0)}{n}}} = \frac{0.66 - 0.50}{\sqrt{\dfrac{0.5(1-0.5)}{1048}}} = \frac{0.16}{0.0154} = 10.39$$

and

$$P\text{-value} = \text{area to the right of 10.39 under a standard normal curve } \leq 0.0002$$

the smallest entry in Table A.

We conclude that this sample provides very strong evidence that more than half of all teens (aged 13 to 17) have a TV in their room.

Exercise 18.35

a) We wish to see if the majority (more than half) of people prefer the taste of fresh-brewed coffee. The hypotheses to be tested are therefore

$$H_0: p = 0.50 \text{ versus } H_a: p > 0.50$$

We see that $n = 50$, $p_0 = 0.50$, $\hat{p} = 31/50 = 0.62$, so the z statistic is

$$z = \frac{\hat{p} - p_0}{\sqrt{\frac{p_0(1-p_0)}{n}}} = \frac{0.62 - 0.50}{\sqrt{\frac{0.50(1-0.50)}{50}}} = \frac{0.12}{\sqrt{0.005}} = 1.70$$

The P-value for this value of the z statistic is the area to the right of 1.70 under a standard normal curve. From Table A this is

$$P\text{-value} = 0.0446$$

Because this is smaller than 0.05, the result is significant at the 5% level.

We may conclude that there is good statistical evidence that the majority (more than half) of people prefer the taste of fresh-brewed coffee. This conclusion is based on data from 50 subjects of which 62% favored the taste of fresh-brewed coffee.

b) From the information given in the problem,

n = sample size = 50

$\hat{p}$ = sample proportion = 0.62

C = level of confidence requested = 0.90

z^* = upper $(1 - C)/2$ critical value of the standard normal distribution = 1.645

Thus the 90% confidence interval for p is

$$\begin{aligned}\hat{p} \pm z^* \sqrt{\frac{\hat{p}(1-\hat{p})}{n}} &= 0.62 \pm 1.645 \sqrt{\frac{0.62(1-0.62)}{50}} \\ &= 0.62 \pm 1.645\sqrt{0.004712} \\ &= 0.62 \pm 0.113\end{aligned}$$

c) You should present the two cups of coffee to the subjects in a random order (i.e., determine which cup a subject gets first by a random mechanism such as flipping a coin).

CHAPTER 19

COMPARING TWO PROPORTIONS

OVERVIEW

Confidence intervals and tests designed to compare two population proportions are based on the **difference in the sample proportions** $\hat{p}_1 - \hat{p}_2$. The formula for the level C confidence interval is

$$(\hat{p}_1 - \hat{p}_2) \pm z^*\text{SE}$$

where z^* is the upper $(1 - C)/2$ standard normal critical value and SE is the standard error for the difference in the two proportions computed as

$$\text{SE} = \sqrt{\frac{\hat{p}_1(1-\hat{p}_1)}{n_1} + \frac{\hat{p}_2(1-\hat{p}_2)}{n_2}}$$

In practice, use this confidence interval when the populations are at least 10 times as large as the samples and the counts of successes and failures are 10 or more in both samples.

To get a more accurate confidence interval for smaller samples., add four imaginary observations, one success and one failure in each sample. Then use the formula above for the confidence interval. This is the **plus four confidence interval.** You can use it whenever both samples have 5 or more observations.

Significance tests for the equality of the two proportions, H_0: $p_1 = p_2$, use a different standard error for the difference in the sample proportions, which is based on a **pooled estimate** of the common (under H_0) value of p_1 and p_2,

$$\hat{p} = \frac{\text{count of successes in both samples combined}}{\text{count of observations in both samples combined}}$$

The test uses the z statistic

$$z = \frac{\hat{p}_1 - \hat{p}_2}{\sqrt{\hat{p}(1-\hat{p})\left(\frac{1}{n_1} + \frac{1}{n_2}\right)}}$$

and P-values are computed using Table A of the standard normal distribution. In practice, use this test when the populations are at least 10 times as large as the samples and the counts of successes and failures are five or more in both samples.

GUIDED SOLUTIONS

Exercise 19.1

KEY CONCEPTS - large sample confidence intervals for the difference between two population proportions

a) The two populations are older black men and older black women. The two sample sizes are

n_1 = number of black men in the sample =

n_2 = number of black women in the sample =

and the number of successes are

number of black men in the sample who said they felt vulnerable =

number of black women in the sample who said they felt vulnerable =

From the data, the estimates of the two proportions are

$\hat{p}_1$ =

$\hat{p}_2$ =

b) Let p_1 represent the proportion of all older black men in Atlantic City, New Jersey who feel vulnerable, p_2 the proportion of all older black women in Atlantic City, New Jersey who feel vulnerable. Recall that a level C confidence interval for $p_1 - p_2$ is

$$(\hat{p}_1 - \hat{p}_2) \pm z^*\text{SE}$$

where z^* is the upper $(1 - C)/2$ standard normal critical value and SE is the standard error for the difference in the two proportions computed as

$$\text{SE} = \sqrt{\frac{\hat{p}_1(1-\hat{p}_1)}{n_1} + \frac{\hat{p}_2(1-\hat{p}_2)}{n_2}}$$

In part (a) you computed $\hat{p}_1$ and $\hat{p}_2$. Now compute the standard error

$$\text{SE} = \sqrt{\frac{\hat{p}_1(1-\hat{p}_1)}{n_1} + \frac{\hat{p}_2(1-\hat{p}_2)}{n_2}} =$$

For a 95% confidence interval,

z^* =

Now compute the interval

$$(\hat{p}_1 - \hat{p}_2) \pm z^*\text{SE} =$$

Exercise 19.16

KEY CONCEPTS - plus four confidence intervals for the difference between two population proportions

a) We can use the large-sample confidence interval when the populations are at least 10 times as large as the samples and the counts of successes and failures are 10 or more in both samples. Are these conditions satisfied?

b) Let p_1 represent the proportion of mice ready to breed in good acorn years, p_2 the proportion in bad acorn years. We should compute the plus four 90% confidence interval. To do this, we must add four imaginary observations, one success and one failure in each sample. When we add these imaginary observations, the number of successes (mice in breeding condition) and the two sample sizes are

number of successes (mice in breeding condition) in first area = ___ + 1 =

n_1 = ___ + 2 =

number of successes (mice in breeding condition) in second area = ___ + 1 =

n_2 = ___ + 2 =

From the data, the plus four estimates of the two proportions are

$\hat{p}_1 =$

$\hat{p}_2 =$

The standard error is

$$\text{SE} = \sqrt{\frac{\hat{p}_1(1-\hat{p}_1)}{n_1} + \frac{\hat{p}_2(1-\hat{p}_2)}{n_2}} =$$

and for a 90% confidence interval

$z^* =$

so our 90% confidence interval for $p_1 - p_2$ is

$$(\hat{p}_1 - \hat{p}_2) \pm z^*\text{SE} =$$

Exercise 19.20

KEY CONCEPTS - testing equality of two population proportions, confidence intervals for the difference between two populations proportions

a) Let p_1 represent the proportion of students from urban/suburban backgrounds who succeed and p_2 the proportion from rural/small-town backgrounds who succeed. Recall that a test of the hypothesis H_0: $p_1 = p_2$ uses the z statistic

$$z = \frac{\hat{p}_1 - \hat{p}_2}{\sqrt{\hat{p}(1-\hat{p})\left(\frac{1}{n_1}+\frac{1}{n_2}\right)}}$$

where n_1 and n_2 are the sizes of the samples, $\hat{p}_1$ and $\hat{p}_2$ the estimates of p_1 and p_2, and

$$\hat{p} = \frac{\text{count of successes in both samples combined}}{\text{count of observations in both samples combined}}$$

First state the hypotheses to be tested . (what is the alternative in this case, one-sided or two-sided?)

The two sample sizes are

$n_1 =$

$n_2 =$

From the data, the estimates of these two proportions are

$\hat{p}_1 =$

$\hat{p}_2 =$

Now compute

$$\hat{p} = \frac{\text{count of successes in both samples combined}}{\text{count of observations in both samples combined}} =$$

and then

$$z = \frac{\hat{p}_1 - \hat{p}_2}{\sqrt{\hat{p}(1-\hat{p})\left(\frac{1}{n_1}+\frac{1}{n_2}\right)}} =$$

Finally, using Table A, compute

P-value =

What do you conclude?

b) Construct the 90% confidence interval, using the steps outlined to guide you.

The two sample sizes are

$n_1 =$

$n_2 =$

Verify that it is safe to use the large-sample confidence interval.

From the data, the estimates of these two proportions are

$\hat{p}_1 =$

$\hat{p}_2 =$

Now compute

$$\text{SE} = \sqrt{\frac{\hat{p}_1(1-\hat{p}_1)}{n_1} + \frac{\hat{p}_2(1-\hat{p}_2)}{n_2}} =$$

For a 90% confidence interval,

$z^* =$

Now compute the interval

$(\hat{p}_1 - \hat{p}_2) \pm z^*\text{SE} =$

Exercise 19.22

KEY CONCEPTS - testing equality of two population proportions

First state the hypotheses to be tested. (what is the alternative in this case; one-sided or two-sided?)

The two sample sizes are

$n_1 =$

$n_2 =$

From the data, the estimates of these two proportions are

$\hat{p}_1 =$

$\hat{p}_2 =$

Now compute

$$\hat{p} = \frac{\text{count of successes in both samples combined}}{\text{count of observations in both samples combined}} =$$

and then

$$z = \frac{\hat{p}_1 - \hat{p}_2}{\sqrt{\hat{p}(1-\hat{p})\left(\frac{1}{n_1} + \frac{1}{n_2}\right)}} =$$

Finally, using Table A, compute

P-value =

What do you conclude?

COMPLETE SOLUTIONS

Exercise 19.1

a) The two sample sizes are

n_1 = number of black men in the sample = 63

n_2 = number of black women in the sample = 56

and the number of successes are

number of black men in the sample who said they felt vulnerable = 46

number of black women in the sample who said they felt vulnerable = 27

From the data, the estimates of the two proportions are

$\hat{p}_1 = 46/63 = 0.73$

$\hat{p}_2 = 27/56 = 0.48$

b) $\text{SE} = \sqrt{\frac{\hat{p}_1(1-\hat{p}_1)}{n_1} + \frac{\hat{p}_2(1-\hat{p}_2)}{n_2}} = \sqrt{\frac{0.73(1-0.73)}{63} + \frac{0.48(1-0.48)}{56}} = \sqrt{0.0075857} = 0.087$

For a 95% confidence interval,

$z^* = 1.96$

so the 95% confidence interval for the difference is

$(\hat{p}_1 - \hat{p}_2) \pm z^*\text{SE} = 0.73 - 0.48 \pm 1.96(0.087) = 0.25 \pm 0.17$

or 0.08 to 0.42.

Exercise 19.16

a) We can use the large-sample confidence interval when the populations are at least 10 times as large as the samples and the counts of successes and failures are 10 or more in both samples. For the 17 mice trapped in the second area, there are 10 successes and 7 failures. Because the number of failures in the second sample is less than 10, the large-sample confidence interval may not be accurate.

b) We compute the plus four 90% confidence interval. We add four imaginary observations, one success and one failure in each sample. When we add these imaginary observations, the number of successes (mice in breeding condition) and the two sample sizes are

number of successes (mice in breeding condition) in first area = 54 + 1 = 55

$n_1 = 72 + 2 = 74$

number of successes (mice in breeding condition) in second area = 10 + 1 = 11

$n_2 = 17 + 2 = 19$

From the data, the plus four estimates of the two proportions are

$$\hat{p}_1 = 55/74 = 0.74$$

$$\hat{p}_2 = 11/19 = 0.58$$

The standard error is

$$SE = \sqrt{\frac{\hat{p}_1(1-\hat{p}_1)}{n_1} + \frac{\hat{p}_2(1-\hat{p}_2)}{n_2}} = \sqrt{\frac{0.74(1-0.74)}{74} + \frac{0.58(1-0.58)}{19}} = \sqrt{0.0026 + 0.0128} = 0.12$$

and for a 90% confidence interval

$$z^* = 1.645$$

so our 90% confidence interval is

$$(\hat{p}_1 - \hat{p}_2) \pm z^*SE = (0.74 - 0.58) \pm 1.645(0.12) = 0.16 \pm 0.20$$

or – 0.04 to 0.36.

Exercise 19.20

a) We are interested in determining whether there is good evidence that the proportion of students who succeed is *different* for urban/suburban versus rural/small-town backgrounds. Thus, the hypotheses to be tested are

$$H_0: p_1 = p_2$$

$$H_a: p_1 \neq p_2$$

The two sample sizes are

n_1 = number from urban/suburban background = 65

n_2 = number from rural/small-town background = 55

From the data, the estimates of the two proportions of students who succeeded are

$$\hat{p}_1 = 52/65 = 0.8$$

$$\hat{p}_2 = 30/55 = 0.545$$

Next we compute

$$\hat{p} = \frac{\text{count of successes in both samples combined}}{\text{count of observations in both samples combined}} = \frac{52+30}{65+55} = 82/120 = 0.683$$

The value of the z-test statistic is

$$z = \frac{\hat{p}_1 - \hat{p}_2}{\sqrt{\hat{p}(1-\hat{p})\left(\frac{1}{n_1}+\frac{1}{n_2}\right)}} = \frac{0.8-0.545}{\sqrt{0.683(1-0.683)\left(\frac{1}{65}+\frac{1}{55}\right)}} = \frac{0.255}{\sqrt{0.00727}} = 2.99$$

Using Table A (we need to double the tail area since this is a two-sided test),

$$P\text{-value} = 2 \times (0.0014) = 0.0028$$

There is good statistical evidence that the proportion of students who succeed is different for urban/suburban versus rural/small-town backgrounds.

b) The two sample sizes are as in part (a), namely,

$$n_1 = 65$$

$$n_2 = 55$$

The number of successes are 52 and 30, and the number of failures are 65 – 52 = 13 and 55 – 30 = 25. All are larger than 10, so we can safely use the large-sample interval.

The estimates of the two proportions are also as in part (a), namely,

$$\hat{p}_1 = 52/65 = 0.8$$

$$\hat{p}_2 = 30/55 = 0.545$$

The standard error is

$$\text{SE} = \sqrt{\frac{\hat{p}_1(1-\hat{p}_1)}{n_1}+\frac{\hat{p}_2(1-\hat{p}_2)}{n_2}} = \sqrt{\frac{0.8(1-0.8)}{65}+\frac{0.545(1-0.545)}{55}} = \sqrt{0.00246+0.00451} = 0.08$$

For a 90% confidence interval,

$$z^* = 1.645$$

so our confidence interval is

$$(\hat{p}_1 - \hat{p}_2) \pm z^*\text{SE} = (0.8 - 0.545) \pm 1.645(0.08) = 0.255 \pm 0.13$$

Exercise 19.22

Let p_1 represent the proportion of female students who will succeed and p_2 the proportion of males who will succeed. We are interested in determining whether there is a difference in these two proportions; hence we should test the hypotheses

$$H_0: p_1 = p_2$$

$$H_a: p_1 \neq p_2$$

The two sample sizes are

n_1 = number of female students in the course = 34

n_2 = number of male students in the course = 89

From the data, the estimates of these two proportions are

$$\hat{p}_1 = 23/34 = 0.6765$$

$$\hat{p}_2 = 60/89 = 0.6742$$

Next we compute

$$\hat{p} = \frac{\text{count of successes in both samples combined}}{\text{count of observations in both samples combined}} = \frac{23+60}{34+89} = 83/123 = 0.6748$$

The value of the z-test statistic is thus

$$z = \frac{\hat{p}_1 - \hat{p}_2}{\sqrt{\hat{p}(1-\hat{p})\left(\frac{1}{n_1}+\frac{1}{n_2}\right)}} = \frac{0.6765-0.6742}{\sqrt{0.6748(1-0.6748)\left(\frac{1}{34}+\frac{1}{89}\right)}} = \frac{0.0023}{\sqrt{0.00892}} = 0.02$$

Finally, we compute the P-value using Table A (we need to double the tail area since this is a two-sided test):

$$P\text{-value} = 2 \times (0.4920) = 0.9840$$

The data provide no real statistical evidence of a difference between the proportion of men and women who succeed.

CHAPTER 20

TWO CATEGORICAL VARIABLES: THE CHI-SQUARE TEST

OVERVIEW

The inference methods of Chapter 8 of your text are first extended to a comparison of more than two population proportions. When comparing more than two proportions, it is best to first do an overall test to see if there is good evidence of any differences among the proportions. Then a detailed follow-up analysis can be performed to decide which proportions are different and to estimate the sizes of the differences.

The overall test for comparing several population proportions arranges the data in a **two-way table**. Two-way tables were introduced in Chapter 6 of your text; the tables are a way of displaying the relationship between any two categorical variables. The tables are also called ***r* × *c* tables**, where ***r*** is the number of rows and ***c*** is the number of columns. Often the rows of the two-way tables correspond to populations or treatment groups, the columns to different categories of the response. When comparing several proportions, there would be only two columns since the response takes only one of two values, and the two columns would represent the successes and failures.

The null hypothesis is $H_0: p_1 = p_2 = \cdots = p_n$ which says that the n population proportions are the same. The alternative is "many sided," as the proportions can differ from each other in a variety of ways. To test this hypothesis, we will compare the **observed counts** with the **expected counts** when H_0 is true. The expected cell counts are computed using the formula

$$\text{expected count} = \frac{\text{row total} \times \text{column total}}{n}$$

where n is the total number of observations.

The statistic we will use to compare the expected counts with the observed counts is the **chi-square statistic**. It measures how far the observed and expected counts are from each other using the formula

$$X^2 = \sum \frac{(\text{observed - expected})^2}{\text{expected}}$$

where we sum up all the $r \times c$ cells in the table.

When the null hypothesis is true, the distribution of the test statistic X^2 is approximately the chi-squared distribution with $(r - 1)(c - 1)$ degrees of freedom. The P-value is the area to the right of X^2 under the chi-square density curve. Use Table E in the back of the book to get the critical values and to compute the P-value. The mean of any chi-squared distribution is equal to its degrees of freedom.

We can use the chi-squared statistic when the data satisfy the following conditions.

- The data are independent SRSs from several populations and each observation is classified according to one categorical variable.

- The data are from a single SRS and each observation is classified according to two categorical variables.
- No more than 20% of the cells in the two-way table have expected counts less than 5.
- All cells have an expected count of at least 1.
- In the special case of the 2 × 2 table, all expected counts should exceed 5 before applying the approximation.

GUIDED SOLUTIONS

Exercise 20.2

KEY CONCEPTS - comparison of more than two proportions.

a) To compute the proportion of successful students in each of the three extracurricular activity groups, it is easiest to first fill in the column totals in the table. Note that in this example, the columns correspond to the "populations" and the rows to successes or failures.

	< 2	2 to 12	> 12
C or better	11	68	3
D or F	9	23	5
Totals			

Now, for each of the three extracurricular activity groups, compute the proportion of successful students (C or better).

Activity group	Proportion of successful students
<2 hours	
2 to 12 hours	
>12 hours	

Complete the bar chart to compare the three proportions of successes.

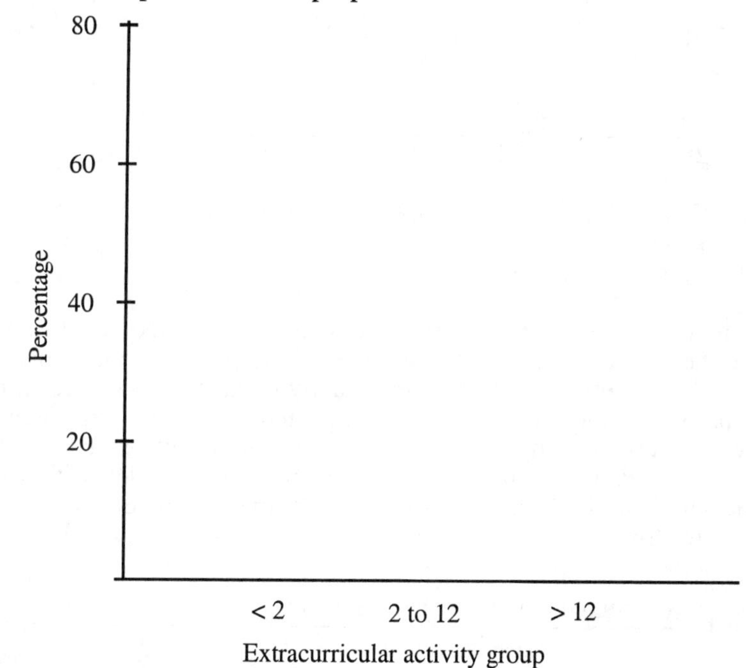

b) What kind of relationship between extracurricular activities and succeeding in the course is shown by these proportions?

Exercise 20.4

KEY CONCEPTS - expected counts

a) Fill in the expected counts in the following table.

	< 2	2 to 12	> 12
C or better	13.78		
D or F			

Remember that the expected cell counts are computed using the formula

$$\text{expected count} = \frac{\text{row total} \times \text{column total}}{n}$$

where n is the total number of observations. The value in the upper left-hand corner was obtained as

$$\text{Expected count} = \frac{\text{row total} \times \text{column total}}{n} = \frac{(82)(20)}{119} = 13.78$$

b) Which cells have the largest deviations between expected and observed counts? What kind of relationship do these suggest?

Exercise 20.6

KEY CONCEPTS - computing the X^2 statistic

(a) We reproduce the Minitab output from Figure 20.4 of your text. Expected counts are printed below observed counts.

	<2	2 to 12	>12	Total
A, B, C	11	68	3	82
	13.78	62.71	5.51	
D or F	9	23	5	37
	6.22	28.29	2.49	
Total	20	91	8	119

Chi-Sq = 0.561 + 0.447 + 1.145 + 1.244 + 0.991 + 2.538 = 6.926
DF = 2, *P*-value = 0.031
1 cells with expected counts less than 5.0

The components of the chi-square statistic are calculated from the expected and observed count in each cell. If you use the expected counts in the table to compute the components of the chi-square statistic, your results will be quite close to but not exactly equal to the values in the Minitab output. This is because the printed output rounds off the expected counts to two decimal places, but the calculations used by Minitab to compute the components of the chi-square statistic are based on "unrounded" values. So if your calculations are based on the "rounded off" values, they will not exactly agree with the values in the Minitab output. Verify the components of the chi-square statistic. We have verified the entry for row 1 and column 1.

Cell in row 1, column 1: $\frac{(\text{observed count - expected count})^2}{\text{expected count}} = \frac{(11 - 13.78)^2}{13.78} = 0.561$

Cell in row 1, column 2:

Cell in row 1, column 3:

Cell in row 2, column 1:

Cell in row 2, column 2:

Cell in row 2, column 3:

We calculate the value of the X^2 statistic by summing these 6 components. The sum is

$$X^2 =$$

b) You can read the *P*-value directly from the output. What is its value, and in simple language, what does it mean to reject H_0 in this setting?

c) Which component of X^2 is the largest? What does this tell you about the relationship between extracurricular activities and academic success?

d) Is this an experiment or an observational study? What would this say about a cause-and-effect relationship?

Exercise 20.8

KEY CONCEPTS - degrees of freedom for the chi-square distribution, *P*-values using Table E

a) What are the values of *r* and *c* in the table? The degrees of freedom can be found using the formula

$$\text{Degrees of freedom} = (r - 1)(c - 1) =$$

b) Go to the row in Table E corresponding to the degrees of freedom found in part a. Between what two entries in Table E does the value $X^2 = 6.926$ lie? What does this tell you about the *P*-value?

Exercise 20.13

KEY CONCEPTS - chi-square test for goodness of fit

a) For the professors grades, fill in the percentages of students earning each grade and compare these to the TA's percentages.

Grade	A	B	C	D/F
Percentage				

b) Fill in the expected counts for each grade. Using the TA percentages as the null probabilities, the expected count for A's is $np_{10} = 91 \times 0.32 = 29.12$.

Grade	A	B	C	D/F
Expected	29.12			

c) Complete the calculation of the chi-square goodness of fit statistic. We have indicated how the first term in the sum should be computed.

$$X^2 = \sum \frac{(\text{count of outcome } i - np_{i0})^2}{np_{i0}} = \frac{(22-29.12)^2}{29.12} +$$

The degrees of freedom are

$k - 1 =$

with k the number of possible outcomes. What can you say about the P-value and what is your conclusion?

Exercise 20.28

KEY CONCEPTS - two-way tables, testing hypotheses with the X^2 statistic

a) Think of the rows of the table (disease status) as the "treatments" and the columns (olive oil consumption) as the response, since this corresponds to how the samples were selected. To be an experiment, what needs to be true about the assignment of the treatments to the subjects? Was this carried out here?

b) The data from the text are reproduced here, with the expected counts printed below the observed counts. Verify a few of the expected counts on your own. If you are not sure how to compute the expected counts, review the complete solution for Exercise 20.4 of this study guide.

	Olive oil			
	Low	Medium	High	Total
Colon cancer	398 404.39	397 404.19	430 416.42	1225
Rectal cancer	250 240.32	241 240.20	237 247.47	728
Controls	1368 1371.29	1377 1370.61	1409 1412.10	4154
Total	2016	2015	2076	6107

Is high olive oil consumption more common among patients without cancer than in patients with colon cancer or rectal cancer? Compute the percentages for each of the three groups on the next page to answer this question.

Group	Percentage with high olive oil consumption
Colon cancer	
Rectal cancer	
Controls	

c) We have given you the expected counts. In the formula for the X^2 statistic, remember that there are 9 terms that need to be summed – one for each cell in the table.

$$X^2 = \sum \frac{(\text{observed - expected})^2}{\text{expected}} =$$

What is the mean of the X^2 statistic under the null hypothesis? If there is evidence to reject the null hypothesis, the computed value of the statistic should be larger than we would expect it to be under the null hypothesis. Is that true in this case?

What is the *P*-value? What do you conclude?

d) If less than 4% of the cases or controls refused to participate, why would our confidence in the results be strengthened?

Complete Solutions

Exercise 20.2

The proportion of successful students in each column are

	< 2	2 to 12	> 12
C or better	11/20 = .550	68/91 = .747	3/8 = .375

and the bar graph of these percentages is

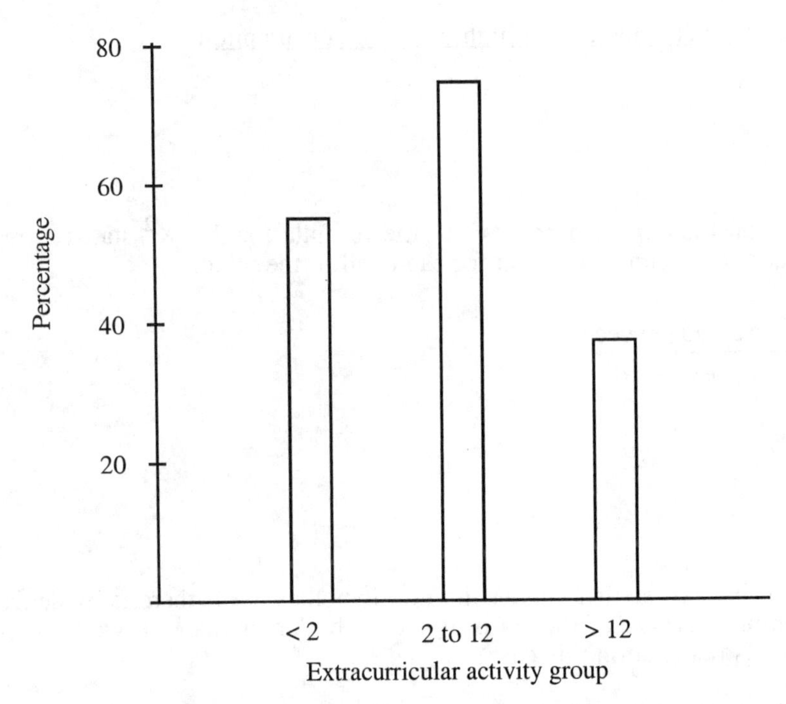

b) The proportions seem to suggest that spending too much time in extracurricular activities is associated with a lack of success in the course. Spending little time in extracurricular activities is more highly associated with success, while spending a moderate amount of time is the group most highly associated with success. Perhaps the maxim "Moderation in all things" applies here!

Exercise 20.4

a) Following is the table of expected counts, with the required calculations.

	< 2	2 to 12	> 12
C or better	13.78	62.71	5.51
D or F	6.22	28.29	2.49

$62.71 = \dfrac{(82)(91)}{119}$ $\quad 5.51 = \dfrac{(82)(8)}{119}$ $\quad 6.22 = \dfrac{(37)(20)}{119}$ $\quad 28.29 = \dfrac{(37)(91)}{119}$ $\quad 2.49 = \dfrac{(37)(8)}{119}$

b) None of the deviations is enormous, but neither is there extremely close agreement between the observed and expected counts. The largest deviations between the observed and expected counts (in an absolute sense) occur for the 2 to 12 hours in extracurricular activities group. The observed number of successful students is larger than expected and the observed number of unsuccessful students is smaller than expected. This may suggest an association between success in schoolwork and a moderate involvement in extracurricular activities, as described in Exercise 20.2 of this study guide.

Exercise 20.6

a) The components of the chi-square statistic are calculated from the expected and observed count in each cell.

Cell in row 1, column 1: $\dfrac{(\text{observed count} - \text{expected count})^2}{\text{expected count}} = \dfrac{(11 - 13.78)^2}{13.78} = 0.561$

Cell in row 1, column 2: $\dfrac{(\text{observed count} - \text{expected count})^2}{\text{expected count}} = \dfrac{(68 - 62.71)^2}{62.71} = 0.446$

Cell in row 1, column 3: $\dfrac{(\text{observed count - expected count})^2}{\text{expected count}} = \dfrac{(3 - 5.51)^2}{5.51} = 1.143$

Cell in row 2, column 1: $\dfrac{(\text{observed count - expected count})^2}{\text{expected count}} = \dfrac{(9 - 6.22)^2}{6.22} = 1.243$

Cell in row 2, column 2: $\dfrac{(\text{observed count - expected count})^2}{\text{expected count}} = \dfrac{(23 - 28.29)^2}{28.29} = 0.989$

Cell in row 2, column 3: $\dfrac{(\text{observed count - expected count})^2}{\text{expected count}} = \dfrac{(5 - 2.49)^2}{2.49} = 2.530$

We calculate the value of the X^2 statistic by summing these 6 components. The sum is

$$X^2 = 6.912$$

b) The *P*-value is the probability of obtaining a value of the X^2 statistic at least as large as the observed value (6.926 on the Minitab output) and is given in the output as 0.031. We would reject H_0 at level 0.05 but not at level 0.01. Rejecting H_0 means that the observed differences in the proportions in each group for the successful ("C or better") and unsuccessful students ("D or F") cannot be easily attributed to chance. In other words, there is evidence that the proportions of successful students in the three groups are different.

c) The term contributing most to X^2 is the term for the number of unsuccessful students spending more than 12 hours per week on extracurricular activities. This points to the fact that if one spends too much time on extracurricular activities, one has little time for schoolwork and is thus likely to be unsuccessful.

d) The study was not a designed experiment and hence does not prove that spending more or less time on extracurricular activity *causes* changes in academic success. It is an observational study. While at first glance it seems plausible that changing the amount of time spent on extracurricular activities ought to cause changes in academic success, this assumes (among other things) that students will trade time spent on schoolwork with time spent on extracurricular activities. However, some students may seek out involvement in extracurricular activities as an escape from schoolwork. If they aren't involved in extracurricular activities, they will simply "goof off." Changes in extracurricular activity thus will not necessarily produce changes in academic performance.

Exercise 20.8

a) There are $r = 2$ rows ("C or better" and "D or F") and $c = 3$ columns ("<2," "2 to 12," and ">12"). Thus the number of degrees of freedom is $(r - 1)(c - 1) = (2 - 1)(3 - 1) = (1)(2) = 2$ which agrees with the value in the Minitab output.

b) If we look in the df = 2 row of Table E, we find the following information.

p	.05	.025
*x**	5.99	7.38

$X^2 = 6.926$ lies between the entries for $p = .05$ and $p = .025$. This tells us that the *P*-value is between .05 and .025.

Exercise 20.13

a) For the professors grades, fill in the percentages of students earning each grade and compare these to the TA's percentages.

Grade	A	B	C	D/F
Percentage	22/91 = 0.24	38/91 = 0.42	20/91 = 0.22	11/91 = .0.12

The professor gave a smaller percentage of A's and higher percentage of D's and F's than the TA. The percentages of B's and C's were similar.

b) Using the TA percentages as the null probabilities, the expected counts for the professor are

$np_{10} = 91 \times 0.32 = 29.12.$

$np_{20} = 91 \times 0.41 = 37.31$

$np_{30} = 91 \times 0.20 = 18.20$

and

$np_{40} = 91 \times 0.07 = 6.37$

Grade	A	B	C	D/F
Expected	29.12	37.31	18.20	6.37

c) The value of the test statistic is

$$X^2 = \frac{(22-29.12)^2}{29.12} + \frac{(38-37.31)^2}{37.31} + \frac{(20-18.20)^2}{18.20} + \frac{(11-6.37)^2}{6.37} = 1.741 + 0.022 + 0.178 + 3.365 = 5.306$$

The degrees of freedom are $k - 1 = 4 - 1 = 3$. Looking in the df = 3 row of Table E, we find the following information

p	.20	.15
x^*	4.64	5.32

giving a P-value between 0.15 and 0.20. There is little evidence that the professor's grade distribution differs from the TA's.

Exercise 20.28

a) We are thinking of the disease groups as the explanatory variable, since the samples were chosen from these groups. The experimenters did not assign the subjects to the disease groups. This makes the data observational, and there is the possibility that differences in olive oil consumption may be a result of confounding variables, not necessarily disease status. (It might be more natural to think of olive oil consumption as the explanatory variable and disease status as the response, but the samples were not selected from the different olive oil consumption groups.)

b) The percentages in each group with high olive oil consumptions are computed in the table. There appears to be very little difference in these percentages, suggesting that olive oil in the diet may be unrelated to these forms of cancer.

Group	Percentage with high olive oil consumption
Colon cancer	430/1225 = 0.351
Rectal cancer	237/728 = 0.326
Controls	1409/4154 = 0.339

c) From the computer output,

$$X^2 = \sum \frac{(\text{observed - expected})^2}{\text{expected}} = 0.101 + 0.128 + 0.443 + 0.390 + 0.003 + 0.443 + 0.008 + 0.030 + 0.007 = 1.552$$

If you used the expected counts in the table, which were rounded to two decimals, your answers may disagree slightly with the computations due to rounding error. The mean of the X^2 statistic is equal to the degrees of freedom, which in this case is $(r - 1)(c - 1) = (3 - 1)(3 - 1) = 4$. Since the numerical value of the X^2 statistic is below the mean, there is little evidence to reject the null hypothesis. As can be seen from the table, the observed and expected counts are quite close. From Table E, you can see that the *P*-value is greater than 0.25. In fact, using statistical software shows the *P*-value = 0.817. These data show no evidence of a relationship between disease and olive oil consumption.

d) If there is a high *nonresponse rate*, then it is possible for the percentages in the table to be quite different than the true percentages due to response bias. In this case, the existence of or lack of a relationship could potentially be due to this bias. However, with 96% of the people responding, the bias if it exists could not be large and would have little impact on the results.

CHAPTER 21

INFERENCE FOR REGRESSION

OVERVIEW

In Chapter 5 of your textbook

we first encountered regression. The assumptions that describe the regression model we use in this chapter are the following.

- We have n observations on an explanatory variable x and a response variable y. Our goal is to study or predict the behavior of y for given values of x.
- For any fixed value of x, the response y varies according to a normal distribution. Repeated responses y are independent of each other.
- The mean response μ_y has a straight-line relationship with x:

$$\mu_y = \alpha + \beta x$$

The slope β and intercept α are unknown parameters.

- The standard deviation of y (call it σ) is the same for all values of x. The value of σ is unknown.

The **true regression line** is $\mu_y = \alpha + \beta x$ and says that the mean response μ_y moves along a straight line as the explanatory variable x changes. The parameters β and α are estimated by the slope b and intercept a of the least-squares regression line, and the formulas for these estimates are

$$b = r\frac{s_y}{s_x}$$

and

$$a = \bar{y} - b\bar{x}$$

where r is the correlation between y and x, $\bar{y}$ is the mean of the y observations, s_y is the standard deviation of the y observations, $\bar{x}$ is the mean of the x observations, and s_x is the standard deviation of the x observations.

The **standard error about the least-squares line** is

$$s = \sqrt{\frac{1}{n-2}\sum \text{residual}^2} = \sqrt{\frac{1}{n-2}\sum (y-\hat{y})^2}$$

where $\hat{y} = a + bx$ is the value we would predict for the response variable based on the least-squares regression line. We use s to estimate the unknown σ in the regression model.

A **level C confidence interval** for β is

$$b \pm t^* \text{SE}_b$$

where t^* is the upper $(1 - C)/2$ critical value for the t distribution with $n - 2$ degrees of freedom and

$$\text{SE}_b = \frac{s}{\sqrt{\sum (x - \bar{x})^2}}$$

is the standard error of the least-squares slope b. SE_b is usually computed using a calculator or statistical software.

The **test of the hypothesis** $H_0\colon \beta = 0$ is based on the t statistic

$$t = \frac{b}{\text{SE}_b}$$

with P-values computed from the t distribution with $n - 2$ degrees of freedom. This test is also a test of the hypothesis that the correlation is 0 in the population.

A **level C confidence interval for the mean response** μ_y when x takes the value x^* is

$$\hat{y} \pm t^* \text{SE}_{\hat{\mu}}$$

where $\hat{y} = a + bx$, t^* is the upper $(1 - C)/2$ critical value for the t distribution with $n - 2$ degrees of freedom and

$$\text{SE}_{\hat{\mu}} = s\sqrt{\frac{1}{n} + \frac{(x^* - \bar{x})^2}{\sum (x - \bar{x})^2}}$$

$\text{SE}_{\hat{\mu}}$ is usually computed using a calculator or statistical software.

A **level C prediction interval for a single observation** on y when x takes the value x^* is

$$\hat{y} \pm t^* \text{SE}_{\hat{y}}$$

where t^* is the upper $(1 - C)/2$ critical value for the t distribution with $n - 2$ degrees of freedom and

$$\text{SE}_{\hat{y}} = s\sqrt{1 + \frac{1}{n} + \frac{(x^* - \bar{x})^2}{\sum (x - \bar{x})^2}}.$$

$\text{SE}_{\hat{y}}$ is usually computed using a calculator or statistical software.

Finally, it is always good practice to check that the data satisfy the linear regression model assumptions before doing inference. Scatterplots and residual plots are useful tools for checking these assumptions.

GUIDED SOLUTIONS

Exercise 21.1

KEY CONCEPTS - scatterplots, correlation, linear regression, residuals, standard error of the least-squares line

a) Sketch your scatterplot on the axes provided below.

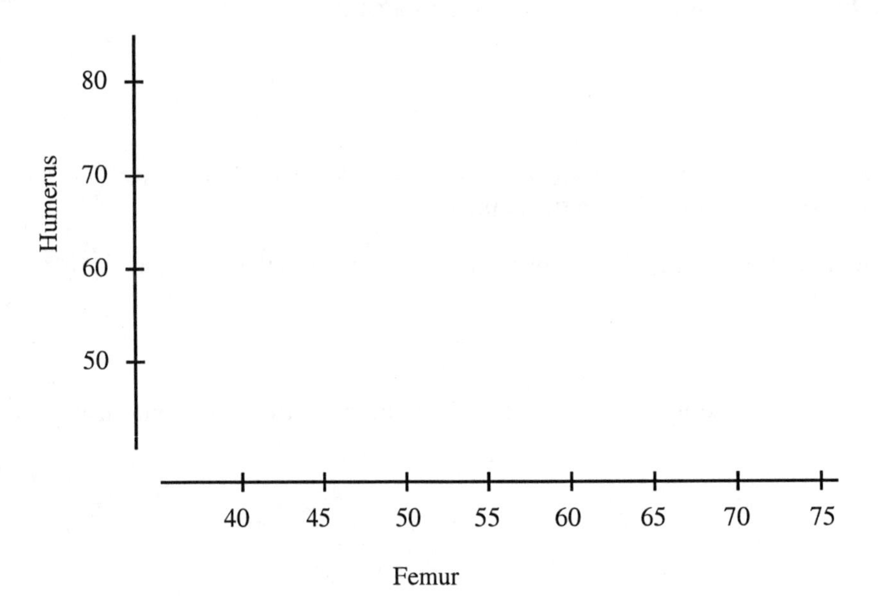

Now use your calculator (or statistical software) to compute the correlation r and the equation of the least-squares regression line.

$r =$

Humerus length =

Do you think femur length will allow a good prediction of humerus length?

b) What does the slope β of the true regression line say about *Archaeopteryx*?

Enter your estimates of the slope β and intercept α of the true regression line in the space provided. Refer to your answer in part (a) for these estimates.

Estimate of $\beta =$

Estimate of $\alpha =$

c) To compute the residuals, complete the table.

Observed value of humerus length	Predicted value of humerus length : −3.65959 + 1.19690(femur length)	Residual (observed − predicted length)
41		
63		
70		
72		
84		

Sum =

Now estimate the standard deviation σ by computing

$$\sum \text{residual}^2 =$$

and then completing the following.

$$s = \sqrt{\frac{1}{n-2}\sum \text{residual}^2} =$$

Exercise 21.6

KEY CONCEPTS - scatterplots, confidence intervals for the slope

a) Use software or the axes provided to make your scatterplot.

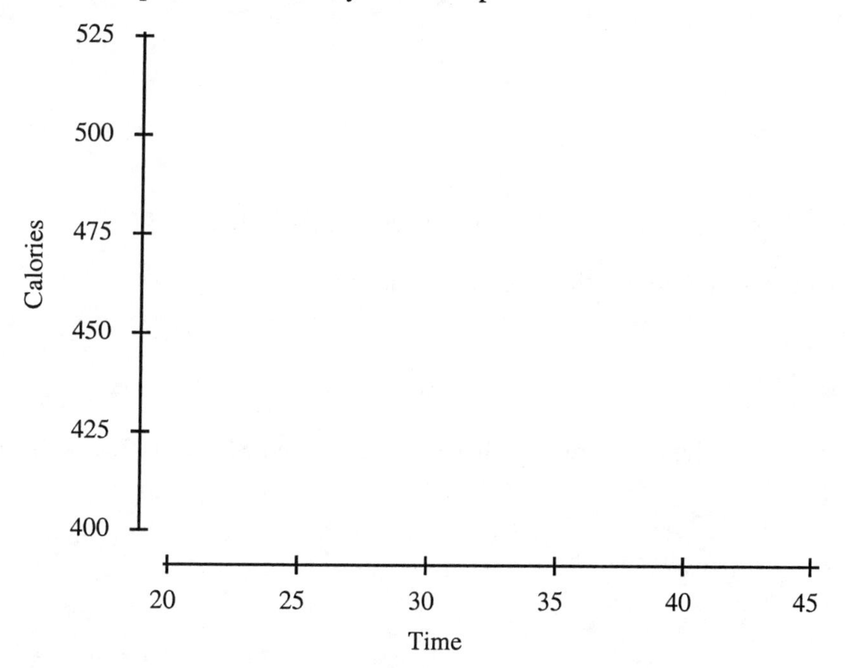

Use software or a calculator to compute the equation of the least-squares regression line. Record the equation below.

$\hat{y} =$

What do the data show about the behavior of children?

To determine the confidence interval, recall that a level C confidence interval for β is

$$b \pm t^* \mathrm{SE}_b$$

where t^* is the upper $(1 - C)/2$ critical value for the t distribution with $n - 2$ degrees of freedom and

$$\mathrm{SE}_b = \frac{s}{\sqrt{\sum (x - \bar{x})^2}}$$

is the standard error of the least-squares slope b. In this exercise b and SE_b can be read directly from the output of statistical software. Record their values below.

$b =$

$\mathrm{SE}_b =$

Now find t^* for a 90% confidence interval from Table C (what is n here?).

$t^* =$

Put all these pieces together to compute the 90% confidence interval.

$b \pm t^* \mathrm{SE}_b =$

Exercise 21.7

KEY CONCEPTS - tests for the slope of the least-squares regression line

a)The test of the hypothesis $H_0{:}\beta = 0$ is based on the t statistic

$$t = \frac{b}{\mathrm{SE}_b}$$

In the statement of the problem, we are told that $b = 1.1969$ and $\mathrm{SE}_b = 0.0751$. Use these values to compute t.

$t = \dfrac{b}{\mathrm{SE}_b} =$

b) What are the degrees of freedom for *t*? Refer to the original data in Exercise 21.1 of your textbook to determine the sample size *n*.

Degrees of freedom = $n - 2$ =

Now use Table C to estimate the *P*-value .

P-value

Exercise 21.14

KEY CONCEPTS - prediction, prediction intervals

We ran Minitab and asked for prediction at Time = 40. The output follows.

```
The regression equation is
Calories = 561 - 3.08 Time

Predictor        Coef         Stdev     t-ratio        p
Constant       560.65         29.37       19.09    0.000
Time          -3.0771        0.8498       -3.62    0.002

s = 23.40         R-sq = 42.1%      R-sq(adj) = 38.9%

Analysis of Variance

SOURCE        DF          SS            MS           F          p
Regression     1      7177.6        7177.6       13.11      0.002
Error         18      9854.4         547.5
Total         19     17032.0

     Fit  Stdev.Fit         95.0% C.I.              95.0% P.I.
  437.57       7.30    (  422.23,   452.91)   (  386.06,   489.08)
```

Where in this output does one find the 95% confidence interval to predict Rachel's calorie consumption at lunch? Refer to Example 21.9 in your textbook if you need help.

95% prediction interval:

Exercise 21.31

KEY CONCEPTS - examining residuals

a) Use software or the axes on the next page to make your residual plot.

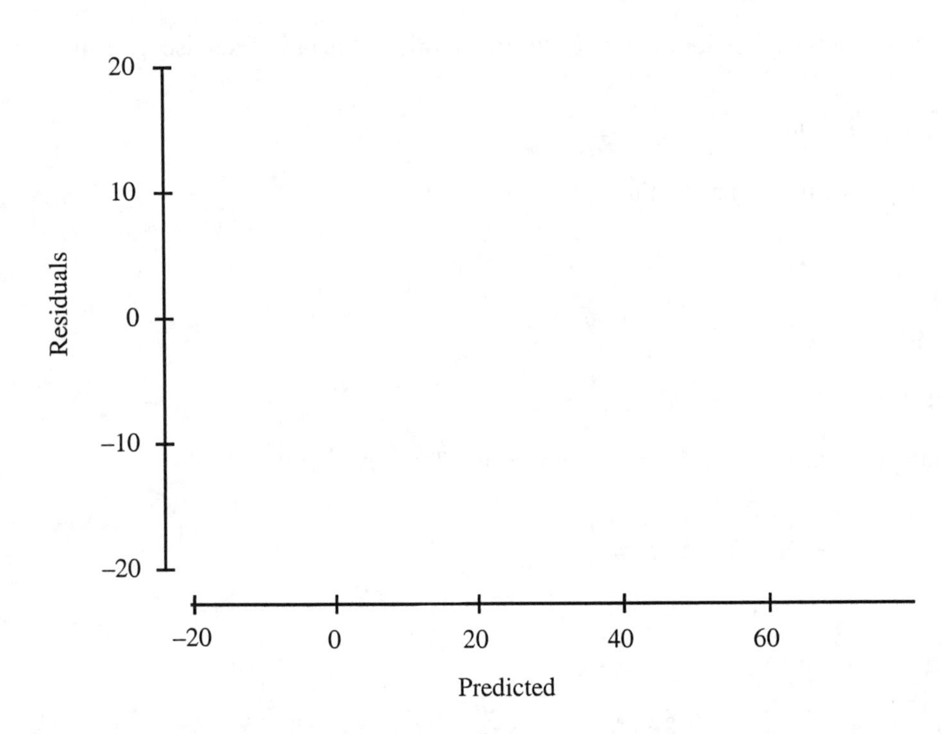

Now make a histogram of the residuals. Use software, or the axes provided below. Use classes beginning with –20 and with intervals of length 5, as suggested by the axes provided.

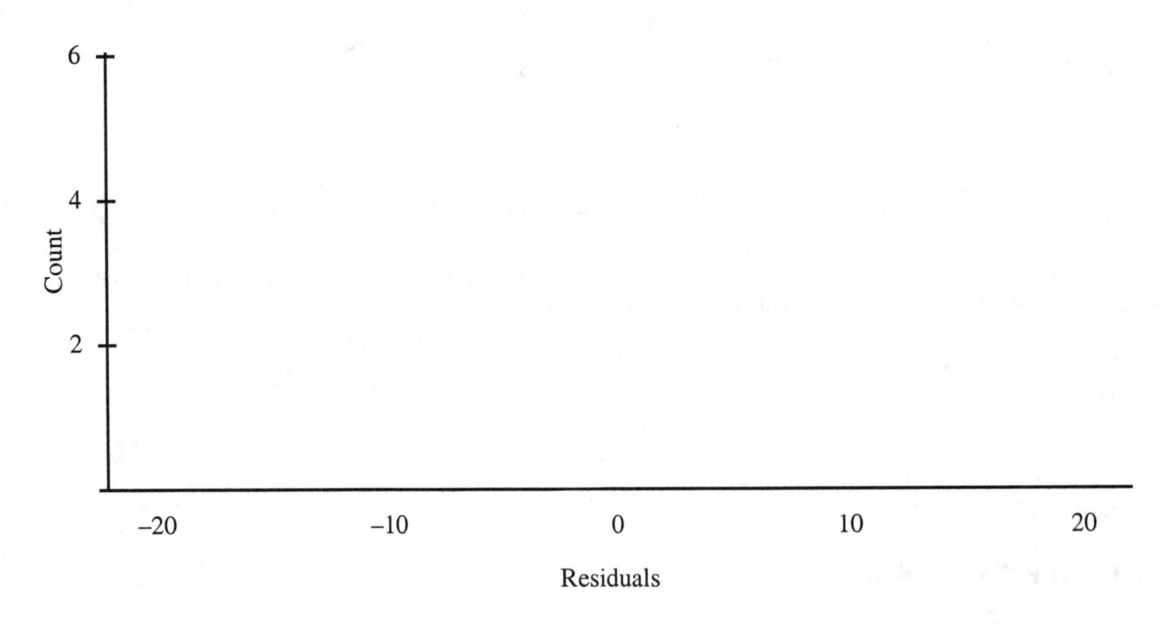

Are there signs of any major violations of the conditions for inference? These conditions are given below. Check each.

- the observations are independent

- the true relationship is linear

• the standard deviation of the response about the true line is the same everywhere

• the response varies Normally about the true regression line

COMPLETE SOLUTIONS

Exercise 21.1

a) If we look at the data, we see that as the length of the femur increases, so does the length of the humerus. Thus there is a positive association between femur and humerus length. A scatterplot of the data with femur length as the explanatory variable follows.

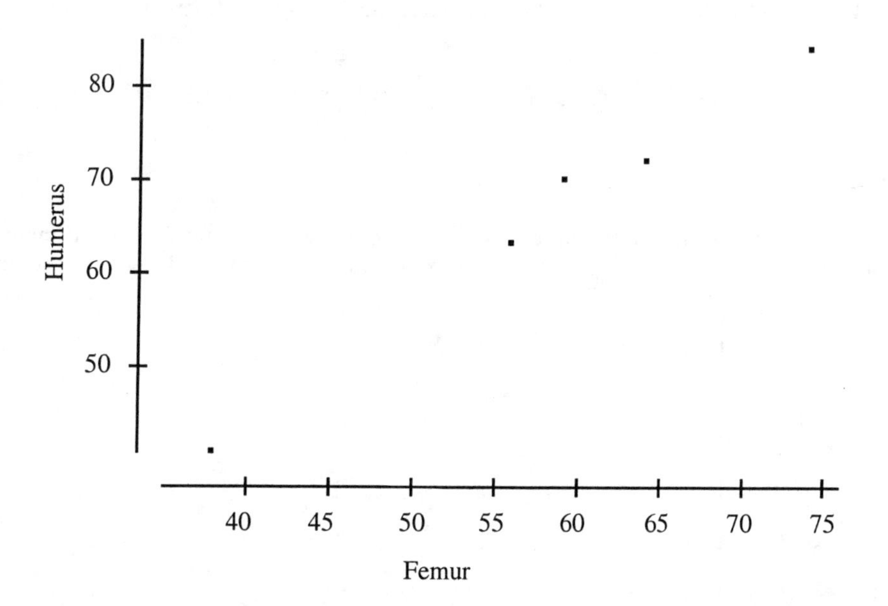

The scatterplot indicates a fairly strong positive association between femur and humerus length. If we calculate the correlation r and the equation of the least-squares line, we obtain the following.

$$r = 0.994$$

$$\text{Humerus length} = -3.65959 + 1.19690 \text{ (femur length)}$$

The correlation is very high, so one would expect that femur length would allow good prediction of humerus length.

b) The slope β of the true regression line tells us the mean increase (in cm) in the length of the humerus associated with a 1-cm increase in the length of the femur in *Archaeopteryx*. From the data,

$$\text{Estimate of } \beta = 1.19690$$

the slope of the least-squares regression line. From the data,

$$\text{Estimate of } \alpha = -3.65959$$

the intercept of the least-squares regression line.

c) The residuals for the five data points are given in the table.

Observed value of humerus length	Predicted value of humerus length : −3.65959 + 1.19690(femur length)	Residual (observed – predicted length)
41	41.822618	−0.822618
63	63.366820	−0.366820
70	66.957520	3.042480
72	72.942021	−0.942021
84	84.911022	−0.911022

The sum of the residuals listed is −.000001, the difference from 0 due to roundoff. To estimate the standard deviation σ in the regression model, we first calculate the sum of the squares of the residuals listed:

$$\sum \text{residual}^2 = 11.785306$$

Our estimate of the standard deviation σ in the regression model is therefore

$$s = \sqrt{\frac{1}{n-2}\sum \text{residual}^2} = \sqrt{\frac{1}{5-2}(11.785306)} = 1.982028$$

Exercise 21.6

Here is a scatterplot showing the relationship between time at the table and calories consumed. Since we are trying to use time to explain calories, time is the explanatory variable and goes on the horizontal axis.

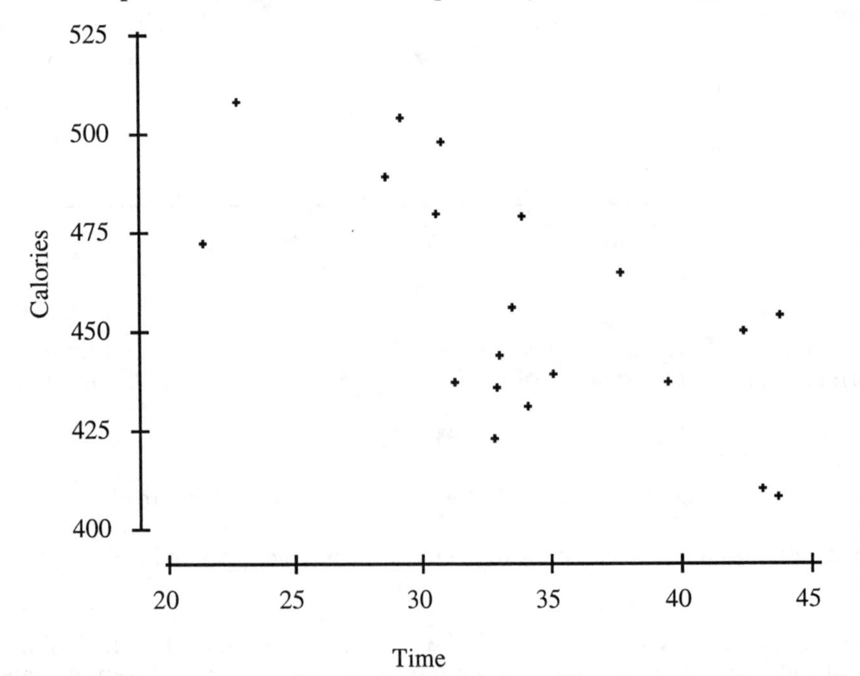

The overall pattern is roughly linear with a negative slope. There are no clear outliers or strongly influential data points.

Using statistical software, we find that the equation of the least-squares line is

$$\hat{y} = 560.65 - 3.08 \times \text{Time}$$

The scatterplot shows that the more time a child spent at the table, the fewer the number of calories the child consumed. The least-squares regression line tells us that each minute increase in time spent at the table is associated with 3.08 fewer calories consumed.

From statistical software we find that

$$b = -3.08$$

$$SE_b = 0.85$$

For a 95% confidence interval from Table C with $n = 20$ (and $n - 2 = 18$),

$$t^* = 2.101$$

We put these pieces together to compute the 95% confidence interval for the true slope of the regression line.

$$b \pm t^* SE_b = -3.08 \pm (2.101)(0.85) = -3.08 \pm 1.79$$

Exercise 21.7

a) $b = 1.1969$ and $SE_b = 0.0751$, so

$$t = \frac{b}{SE_b} = \frac{1.1969}{0.0751} = 15.94$$

b) Referring to the original data in Exercise 21.1 of your textbook we see that $n = 5$

Degrees of freedom $= n - 2 = 5 - 2 = 3$

We use Table C to estimate the P-value . We look in the row corresponding to 3 df. We see that the largest entry in the row is 12.92. Because our t is larger than this we conclude

P-value < 0.0005

Exercise 21.14

The output from Minitab is as follows.

```
The regression equation is
Calories = 561 - 3.08 Time

Predictor        Coef        Stdev     t-ratio        p
Constant       560.65        29.37       19.09    0.000
Time          -3.0771       0.8498       -3.62    0.002

s = 23.40        R-sq = 42.1%      R-sq(adj) = 38.9%

Analysis of Variance

SOURCE        DF          SS          MS         F        p
Regression     1      7177.6      7177.6     13.11    0.002
Error         18      9854.4       547.5
Total         19     17032.0

     Fit  Stdev.Fit        95.0% C.I.           95.0% P.I.
  437.57       7.30    (422.23,   452.91)    (386.06,   489.08)
```

The "Fit" entry gives the predicted calories. Minitab gives both the 95% confidence interval for the mean response and prediction interval for a single observation. We are predicting a single observation, so the column labeled "95% PI" contains the interval we want. We see

95% prediction interval: (386.06, 489.08)

Exercise 21.31

Here is the plot of the residuals against the predicted return.

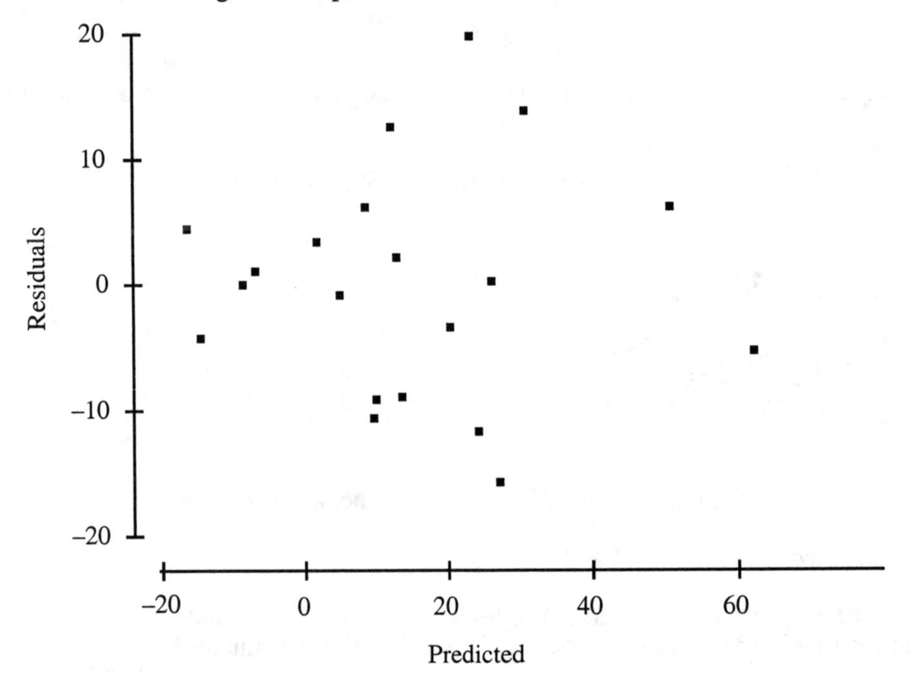

Here is the histogram of the residuals.

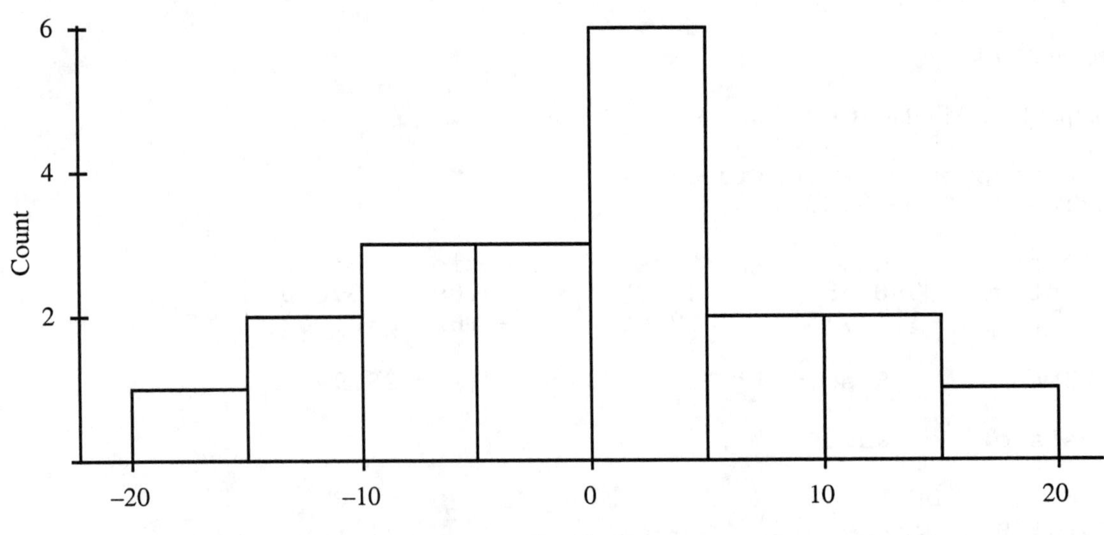

We check each of the conditions.

• The observations are independent: There is no evidence of lack of independence

• The true relationship is linear: The residuals show no trends. There is no evidence that the true relationship is not linear.

• The standard deviation of the response about the true line is the same everywhere: The residuals appear to be scattered more widely about predicted values greater than 5 than predicted values below 5, but there are only a few residuals below 5 so it is difficult to know if this pattern is real. We would argue that there is not strong evidence that the standard deviation of the response about the true line is not the same everywhere.

• The response varies Normally about the true regression line: The histogram shows no significant skewness. There is not strong evidence that the response does not vary Normally about the true regression line.

CHAPTER 22

ONE-WAY ANALYSIS OF VARIANCE: COMPARING SEVERAL MEANS

OVERVIEW

The two-sample t procedures compare the means of two populations. However, when the mean is the best description of the center of a distribution, we may want to compare several population means or several treatment means in a designed experiment. For example, we might be interested in comparing the mean weight loss by dieters on three different diet programs or the mean yield of four varieties of green beans.

The method we use to compare more than two population means is the **analysis of variance (ANOVA) *F* test**. This test is also called the **one-way ANOVA**. The ANOVA F test is an overall test that looks for any difference between a group of I means. The null hypothesis is H_0: $\mu_1 = \mu_2 = \ldots = \mu_I$, where we tell the population means apart by using the subscripts 1 through I. The alternative hypothesis is H_a: not all the means are equal. In a more advanced course, you would study formal inference procedures for a follow-up analysis to decide which means differ and to estimate how large the differences are. Note that formally the ANOVA F test is a different test from the F test you may have studied in Chapter 17 of your text that compared the standard deviations of two populations, although the ANOVA F test does involve the comparison of two measures of variation.

The ANOVA F test compares the variation among the groups to the variation within the groups through the ***F* statistic**,

$$F = \frac{\text{variation among the sample means}}{\text{variation among the individuals in the same sample}}$$

The important thing to take away from this chapter is the rationale behind the ANOVA F test. The particulars of the calculation are not as important since software usually calculates the numbers for us.

The F statistic has the F distribution. The distribution is completely defined by its two degrees of freedom parameters, the numerator degrees of freedom and the denominator degrees of freedom. The numerator has $I - 1$ degrees of freedom, where I is the number of populations we are comparing. The denominator has $N - I$ degrees of freedom, where N is the total number of observations. The F distribution is usually written $F(I - 1, N - I)$.

The assumptions for ANOVA are that

- There are I independent SRSs.
- Each population is normally distributed with its own mean, μ_i.
- All populations have the same standard deviation, σ.

The first assumption is the most important. The test is robust against nonnormality, but it is still important to check for outliers and/or skewness that would make the mean a poor measure of the center of the distribution. As for the assumption of equal standard deviations, make sure that the largest sample standard deviation is no more than twice the smallest standard deviation.

Although it is generally best to leave the ANOVA computations to statistical software, seeing the formulas sometimes helps one to obtain a better understanding of the procedure. In addition, there are times when the original data are not available, and you have only the group means and standard deviations or standard error. In these instances, the formulas described here are required to carry out the ANOVA *F* test.

The *F* statistic is $F = \dfrac{\text{MSG}}{\text{MSE}}$, where MSG is the **mean square for groups,**

$$\text{MSG} = \frac{n_1(\bar{x}_1 - \bar{x})^2 + n_2(\bar{x}_2 - \bar{x})^2 + \ . \ . \ . \ + n_I(\bar{x}_I - \bar{x})^2}{I - 1}$$

with

$$\bar{x} = \frac{n_1\bar{x}_1 + n_2\bar{x}_2 + \ldots + n_I\bar{x}_I}{N}$$

and MSE is the **error mean square,**

$$\text{MSE} = \frac{s_1^2(n_1 - 1) + s_2^2(n_2 - 1) + . \ . \ . + s_I^2(n_I - 1)}{N - I}.$$

Because MSE is an average of the individual sample variances, it is also called the **pooled sample variance,** written s_p^2, and its square root, $s_p = \sqrt{\text{MSE}}$ is called the **pooled standard deviation**. We can also make a confidence interval for any of the means by using the formula $\bar{x}_i \pm t^* \dfrac{s_p}{\sqrt{n_i}}$. The critical value is t^* from the t distribution with $N - I$ degrees of freedom.

GUIDED SOLUTIONS

Exercise 22.3

KEY CONCEPTS - side-by-side stemplots, ANOVA hypotheses, drawing conclusions from ANOVA output

a) Complete the stemplots (they use split stems). From the stemplots, would you say that any of the groups show outliers or extreme skewness? What effects of logging are visible from the stemplots?

Never logged	Logged 1 year ago	Logged 8 years ago
0 |	0 |	0 |
0 |	0 |	0 |
1 |	1 |	1 |
1 |	1 |	1 |
2 |	2 |	2 |
2 |	2 |	2 |
3 |	3 |	3 |

b) What do the means suggest about the effect of logging?

c) State the null and alternative hypotheses, letting μ_1, μ_2 and μ_3 denote the means for the three groups.

H_0: H_a:

From the output, determine the values of the ANOVA F statistic and its P-value. What are your conclusions?

F statistic = P-value =

Exercise 22.5

KEY CONCEPTS - ANOVA degrees of freedom, computing P-values from Table D

a) In the table, fill in the numerical values and explain in words the meaning of each symbol we are using in the notation for the one-way ANOVA. Group 1, group 2 and group 3 are identified in the exercise.

Symbol	Value	Verbal meaning
I		
n_1		
n_2		
n_3		
N		

b) Use the text formulas and the results from part (a) to give the numerator and denominator degrees of freedom. Check your answers against the Excel output.

Numerator degrees of freedom =

Denominator degrees of freedom =

c) The value $F = 11.43$ needs to be referred to an $F(2, 30)$ distribution. What can you say about the P-value from Table D?

Exercise 22.10

KEY CONCEPTS - comparison between means, checking assumptions, interpreting results from an ANOVA

a) By comparing the group means, what is the relationship between marital status and salary?

b) The ratio of the largest to the smallest standard deviations is

$$\frac{\text{largest sample standard deviation}}{\text{smallest sample standard deviation}} =$$

Does this allow the use of the ANOVA F test?

c) Calculate the degrees of freedom for the ANOVA F test by first computing N and I.

$N =$ $I =$

Numerator df = Denominator df =

d) The large sample sizes (particularly for the married men) indicate that the margins of error for the sample means will be very small, much smaller than the observed differences in the means. What is the numerical value of the standard error for married men? How does it compare to the differences in mean salaries between married men and the other groups?

e) Is this an observational study or an experiment? How does that affect the type of conclusions that can be made?

Exercise 22.13

KEY CONCEPTS - ANOVA computations, standard errors

a) The data are given in the form "mean ± standard error." The computation of the ANOVA F statistic requires the three group means and standard deviations. The means are given directly, and the standard deviations can be derived easily from the standard error (SE) by recalling that

$$\mathrm{SE} = \frac{s}{\sqrt{n}}, \text{ or equivalently } s = \mathrm{SE}\sqrt{n}$$

To simplify the calculations that need to be done, complete the following table. We have completed the first line for you, where we are given that $n_1 = 16$ eggs hatched at the cold temperature, $\bar{x}_1 = 28.89$, $\mathrm{SE}_1 = 8.08$, and we compute

$$s_1 = \mathrm{SE}_1\sqrt{n_1} = 8.08\sqrt{16} = 32.32$$

Temperature	$\bar{x}_i$	SE_i	n_i	s_i
Cold	28.89	8.08	16	32.32
Neutral				
Hot				

Do the standard deviations satisfy the rule of thumb for using ANOVA?

$$\frac{\text{largest sample standard deviation}}{\text{smallest sample standard deviation}} =$$

b) You will need the means, sample sizes and standard deviations from your table in part (a) to do the calculations. To compute MSG, you first need to compute the overall mean

$$\bar{x} = \frac{n_1\bar{x}_1 + n_2\bar{x}_2 + \ldots + n_I\bar{x}_I}{N} =$$

and then substitute the means, sample sizes, and overall mean into the formula

$$\text{MSG} = \frac{n_1(\bar{x}_1 - \bar{x})^2 + n_2(\bar{x}_2 - \bar{x})^2 + \ . \ . \ . \ + n_I(\bar{x}_I - \bar{x})^2}{I - 1} =$$

MSE is then obtained from the formula

$$\text{MSE} = \frac{s_1^2(n_1 - 1) + s_2^2(n_2 - 1) + . \ . \ . + s_I^2(n_I - 1)}{N - I} =$$

Finally,

$$F = \frac{\text{MSG}}{\text{MSE}} =$$

What are the degrees of freedom for the ANOVA F statistic? Compare the value you calculated to the critical values in Table D. What is the P-value and is there evidence that nest temperature affects the mean weight of newly hatched pythons?

COMPLETE SOLUTIONS

Exercise 22.3

a) The side-by-side stemplots are completed here. Extreme skewness is not evident. There is a low outlier in the "Logged 8 years ago" column. The counts of trees in the plots that were never logged appear to be larger than those that were logged in the stemplots. There appears to be little difference between the counts from plots logged 1 year ago and plots logged 8 years ago.

Never logged		Logged 1 year ago		Logged 8 years ago	
0		0	2	0	4
0		0	9	0	
1		1	2 2 4 4	1	2 2
1	6 9 9	1	5 7 7 8 9	1	5 8 8 9
2	0 1 2 4	2	0	2	2 2
2	7 7 8 9	2		2	
3	3	3		3	

b) The means suggest that logging reduces the number of trees per plot and that the recovery may be quite slow as there is little difference in the means for the logged 1 year ago and logged 8 years ago groups.

c) The hypotheses are $H_0: \mu_1 = \mu_2 = \mu_3$ and H_a: not all of μ_1, μ_2 and μ_3 are equal.

The overall ANOVA F test has $F = 11.4257$ and P-value = 0.000205, so there is strong evidence of a difference in mean trees per plot among the three groups. The ANOVA F test does not tell us which groups are different, but examination of the means and stemplots shows both the logged groups to have lower mean trees per plot than the never logged group, but there appears to be little difference between the plots logged 1 year ago and the plots logged 8 years ago.

Exercise 22.5

a) The table gives the value and states in words the meaning of each symbol used in a one-way ANOVA.

Symbol	Value	Verbal meaning
I	3	Number of groups
n_1	12	Number of plots in the never logged group
n_2	12	Number of plots in the logged 1 year ago group
n_3	9	Number of plots in the logged 8 years ago group
N	33	Total number of plots in the experiment

b) The ANOVA F statistic has the F distribution with $I - 1 = 3 - 1 = 2$ degrees of freedom in the numerator and $N - I = 33 - 3 = 30$ degrees of freedom in the denominator.

c) The critical value of 9.22 corresponds to a tail probability of 0.001 for an $F(2, 25)$ distribution. Since the critical value for an $F(2, 30)$ distribution corresponding to a tail area of 0.001 would be even smaller, the value $F = 11.4257$ would exceed this value. We can say that the P-value is less than 0.001, which agrees with the computer output.

Exercise 22.10

a) The most obvious feature is that men who are or have been married earn more, on average, than single men. Men who are or have been married earn about the same amount, although divorced men appear to earn a little less, on average, then married or widowed men.

b) The ratio of the largest to the smallest standard deviations is

$$\frac{\text{largest sample standard deviation}}{\text{smallest sample standard deviation}} = \frac{8119}{5731} = 1.42 < 2$$

Since this ratio is less than 2, the sample standard deviations allow the use of the ANOVA F test.

c) We calculate

I = number of populations we wish to compare
= number of different marital statuses
= 4

n_1 = sample size from population 1 (single men) = 337
n_2 = sample size from population 2 (married men) = 7730
n_3 = sample size from population 3 (divorced men) = 126
n_4 = sample size from population 4 (widowed men) = 42

N = total sample size = sum of the n_i = 8235.

The degrees of freedom for the ANOVA F statistic are

$$I - 1 = 4 - 1 = 3$$

degrees of freedom in the numerator and

$$N - I = 8235 - 4 = 8231$$

degrees of freedom in the denominator.

d) The large sample sizes (particularly for the married men) indicate that the margins of error for the sample means will be very small, much smaller than the observed differences in the means. For example, the standard error for the mean of married men is

$$\frac{\$7159}{\sqrt{7730}} = \$81.4$$

which is very small compared with the difference of, for example, over $5000 in mean salaries with single men (you can check that the standard error for single men is $312.2, also much smaller than the difference in means of over $5000).

e) The differences in means probably do not mean that getting married raises men's mean incomes. This is an observational study, so it is not safe to conclude that the observed differences are due to cause and effect. A likely explanation of the observed differences is that the typical single man is much younger than the typical married man. As men get older they are more likely to be married. Younger men have been in the firm for less time than older men and so will have lower salaries. This explanation may also explain the small differences between men who are or have been married. Married and widowed men (as compared to divorced men) are likely to include the most senior men in the firm. The most senior men are likely to have the highest salaries.

Exercise 22.13

a) The completed table follows.

Temperature	$\bar{x}_i$	SE_i	n_i	s_i
Cold	28.89	8.08	16	32.320
Neutral	32.93	5.61	75	48.584
Hot	32.27	4.10	38	25.274

The ratio of the largest to the smallest standard deviations is

$$\frac{\text{largest sample standard deviation}}{\text{smallest sample standard deviation}} = \frac{48.584}{25.274} = 1.92 < 2$$

so the rule of thumb is satisfied.

b) Using the means, sample sizes, and standard deviations from the table in part (a)

$$\bar{x} = \frac{n_1\bar{x}_1 + n_2\bar{x}_2 + \ldots + n_I\bar{x}_I}{N} = \frac{(16)(28.89)+)(75)(32.93)+(38)(32.27)}{129} = 32.234$$

since $N = 16 + 75 + 38 = 129$.

Substituting into the formula for MSG,

$$\text{MSG} = \frac{n_1(\bar{x}_1 - \bar{x})^2 + n_2(\bar{x}_2 - \bar{x})^2 + \ . \ . \ . \ + n_I(\bar{x}_I - \bar{x})^2}{I\text{-}1} =$$

$$= \frac{16(28.89 - 32.234)^2 + 75(32.93 - 32.234)^2 + 38(32.27 - 32.234)^2}{3\text{-}1}$$

$$= 107.649$$

To complete the calculations, we have

$$\text{MSE} = \frac{s_1^2(n_1 - 1) + s_2^2(n_2 - 1) + . \ . \ . + s_I^2(n_I - 1)}{N - I}$$

$$= \frac{32.320^2(15) + 48.584^2(74) + 25.274^2(37)}{126} = 1698.201$$

Finally,

$$F = \frac{\text{MSG}}{\text{MSE}} = \frac{107.649}{1698.201} = 0.063$$

The F-value should be compared to critical values from the $F(2,\ 126)$ distribution or, being conservative, to the $F(2,\ 100)$ distribution. The 0.10 critical value is 2.36, so there is no statistical evidence that nest temperature affects the mean weight of newly hatched pythons.

CHAPTER 23

NONPARAMETRIC TESTS

OVERVIEW

Many of the statistical procedures described in previous chapters assumed that the samples were drawn from normal populations. **Nonparametric tests** do not require any specific form for the distributions of the populations from which the samples were drawn. Many nonparametric tests are **rank tests**; that is, they are based on the **ranks** of the observations rather than on the observations themselves. When ranking the observations from smallest to largest, tied observations receive the average of their ranks.

The **Wilcoxon rank sum test** compares two distributions. The objective is to determine if one distribution has systematically larger values than the other. The observations are ranked, and the **Wilcoxon rank sum statistic** W is the sum of the ranks of one of the samples. The Wilcoxon rank sum test can be used in place of the **two-sample** t **test** when samples are small or the populations are far from normal.

Exact P-values require special tables and are produced by some statistical software. However, many statistical software packages give only approximate P-values based on a normal approximation, typically with a continuity correction employed. Many packages also make an adjustment in the normal approximation when there are ties in the ranks.

The **Wilcoxon signed rank test** is a nonparametric test for matched pairs. It tests the null hypothesis that there is no systematic difference between the observations within a pair against the alternative that one observation tends to be larger.

The test is based on the **Wilcoxon signed rank statistic** W^+, which provides another example of a nonparametric test using ranks. The absolute values of the differences between matched pairs of observations are ranked and the sum of the ranks of the positive (or negative) differences gives the value of W^+. The **matched pairs** t **test** is an alternative test that assumes a normal distribution for the differences.

P-values can be found from special tables of the distribution or a normal approximation to the distribution of W^+. Some software computes the exact P-value and other software uses the normal approximation, typically with a ties correction. Many packages make an adjustment in the normal approximation when there are ties in the ranks.

The **Kruskal-Wallis test** is the nonparametric test for the **one-way analysis of variance** setting. In comparing several populations, it tests the null hypothesis that the distribution of the response variable is the same in all groups and the alternative hypothesis that some groups have distributions of the response variable that are systematically larger than others.

The **Kruskal-Wallis statistic** H compares the average ranks received for the different samples. If the alternative is true, some of these should be larger than others. Computationally, it essentially arises from performing the usual one-way ANOVA to the ranks of the observations rather than the observations themselves.

P-values can be found from special tables of the distribution or a chi-square approximation to the distribution of H. When the sample sizes are not too small, the distribution of H for comparing I populations has approximately a chi-square distribution with $I - 1$ degrees of freedom. Some software computes the exact P-value and other software uses the chi-square approximation, typically with an adjustment in the chi-square approximation when there are ties in the ranks.

GUIDED SOLUTIONS

Exercise 23.13

KEY CONCEPTS - ranking data, two-sample problem, Wilcoxon rank sum test

a) Order the observations from smallest to largest in the space provided. Use a different color or underline those observations in the supplemented group. This will make it easier to determine the ranks assigned to each group.

b) Now suppose the first sample is the supplemented group and the second sample is the control group. The choice of which sample we call the first sample and which we call the second sample is arbitrary. However, the Wilcoxon rank sum test is the sum of the ranks of the first sample, and the formulas for the mean and variance of W distinguish between the sample sizes for the first and the second samples. Use the ranks of the supplemented group to compute the value of W.

$W =$

c) What are the values of n_1, n_2 and N? Use these to evaluate the mean and standard deviation of W according to the formulas below.

$$\mu_W = \frac{n_1(N+1)}{2} =$$

$$\sigma_W = \sqrt{\frac{n_1 n_2 (N+1)}{12}} =$$

Now use the mean and standard deviation to compute the standardized rank sum statistic

$$z = \frac{W - \mu_W}{\sigma_W} =$$

What kind of values would W have if the alternative were true? Use the normal approximation to find the approximate P-value. If you have access to software or tables to evaluate the exact P-value, compare it with the approximation.

What are your conclusions?

Exercise 23.29

KEY CONCEPTS - matched pairs, Wilcoxon signed rank statistic

a) First give the null and alternative hypotheses. If the cola loses sweetness, what will be the sign of the sweetness loss (sweetness before storage minus sweetness after storage)?

H_0:

H_a:

To compute the Wilcoxon signed rank statistic, first order the absolute values of the differences and rank them. When there are ties, you need to be careful computing the ranks. For any tied group of observations, they should each receive the average rank for the group. (Note that the negative observations are in bold and italics.) The ranks of the two smallest absolute values are given to help get you started. Now fill in the remaining ranks.

Absolute values	Ranks
0.4	1.5
0.4	1.5
0.7	
1.1	
1.2	
1.3	
2.0	
2.0	
2.2	
2.3	

To see how the ranks are computed, the 0.4's would get ranks 1 and 2 so their average rank is 1.5. The 0.7 would get rank 3, and so on. If W^+ is the sum of the ranks of the positive observations, compute the value of W^+.

$W^+ =$

Evaluate the mean and standard deviation of W^+ according to the formulas below.

$$\mu_{W^+} = \frac{n(n+1)}{4}$$

$$\sigma_{W^+} = \sqrt{\frac{n(n+1)(2n+1)}{24}}$$

Now use the mean and standard deviation to compute the standardized rank sum statistic

$$z = \frac{W^+ - \mu_{W^+}}{\sigma_{W^+}} =$$

Do you expect W^+ to be small or large if the alternative is true? Use the normal approximation to find the approximate P-value.

What are your conclusions?

b) How do the P-values from the Wilcoxon signed rank test and the one-sample t test compare?

For the one-sample t test, give the null and alternative hypotheses.

H_0:

H_a:

What are the assumptions for each of the procedures?

Exercise 23.39

KEY CONCEPTS - one-way ANOVA, Kruskal-Wallis statistic

a) Think carefully about the differences in the hypotheses being tested.

ANOVA test:

H_0:

H_a:

Kruskal-Wallis test:

H_0:

H_a:

b) Find the median for each group. Recall from Chapter 2 of your text, that the median is the middle observation after they have been ordered. When the sample sizes are even, the median is the average of the two middle observations.

Nematodes	Median
0	
1000	
5000	
10000	

Using the information in the medians, do the nematodes appear to retard growth?

To compute the Kruskal-Wallis test statistic, the 16 observations are first arranged in increasing order. That step has been carried out below, where we have kept track of the group for each observation. You need to fill in the ranks in the line provided. Remember that there is one tied observation.

Growth	3.2	4.6	5.0	5.3	5.4	5.8	7.4
Group	10000	5000	5000	10000	5000	10000	5000
Rank							

Growth	7.5	8.2	9.1	9.2	10.8	11.1	11.1
Group	10000	1000	0	0	0	1000	1000
Rank							

Growth	11.3	13.5
Group	1000	0
Rank		

Now fill in the table below which gives the ranks for each of the nematode groups, and the sum of ranks for each group.

Nematodes	Ranks	Sum of Ranks
0		
1000		
5000		
10000		

Use the sum of ranks for the four groups to evaluate the Kruskal-Wallis statistic. What are the numerical values of the n_i and N in the formula?

$$H = \frac{12}{N(N+1)} \sum \frac{R_i^2}{n_i} - 3(N+1) \quad =$$

The value of H is compared with critical values in Table E for a chi-square distribution with $I - 1$ degrees of freedom, where I is the number of groups. What is the P-value and what do you conclude?

COMPLETE SOLUTIONS

Exercise 23.13

a) The observations are first ordered from smallest to largest. The observations in bold are from the supplemented group.

Observations	Ranks
−1.2	1
2.3	2
4.6	3.5
4.6	3.5
5.4	5
6.0	6
7.7	7.5
7.7	7.5
11.3	9.5
11.3	9.5
11.4	11
15.5	12
16.5	13

b) The Wilcoxon rank sum statistic is

$W = 5 + 7.5 + 9.5 + 9.5 + 11 + 12 + 13 = 67.5$

c) The sample sizes are $n_1 = 7$, $n_2 = 6$, and $N = 13$. The values for the mean and variance are

$$\mu_W = \frac{n_1(N+1)}{2} = \frac{7(13)}{2} = 45.5 \text{ and}$$

$$\sigma_W = \sqrt{\frac{n_1 n_2 (N+1)}{12}} = \sqrt{\frac{(7)(6)(13)}{12}} = 6.745$$

and the standardized rank sum statistic W is

$$z = \frac{W - \mu_W}{\sigma_W} = \frac{67.5 - 45.5}{6.745} = 3.26$$

Since we would expect W to have large values if the alternative were true, the approximate P-value is $P(Z \geq 3.26) = 0.0006$. There is very strong evidence that the supplemented birds miss the peak by more days than the control birds.

Exercise 23.29

a) The null and alternative hypotheses are

H_0: median = 0

H_a: median > 0.

The ranks of the absolute values are

Absolute values	Ranks
0.4	1.5
0.4	1.5
0.7	3
1.1	4
1.2	5
1.3	6
2.0	7.5
2.0	7.5
2.2	9
2.3	10

The Wilcoxon signed rank statistic is

$W^+ = 1.5 + 3 + 4 + 5 + 7.5 + 7.5 + 9 + 10 = 47.5$

The values for the mean and variance are

$$\mu_{W^+} = \frac{n(n+1)}{4} = \frac{10(11)}{4} = 27.5 \text{ and}$$

$$\sigma_{W^+} = \sqrt{\frac{n(n+1)(2n+1)}{24}} = \sqrt{\frac{(10)(11)(21)}{24}} = 9.811$$

and the standardized signed rank statistic W^+ is

$$\frac{W^+ - \mu_{W^+}}{\sigma_{W^+}} \geq \frac{47.5 - 27.5}{9.811} = 2.04$$

If the cola lost sweetness, we would expect the differences (before storage – after storage) to be positive. Thus the ranks of the positive observations should be large and we would expect the value of the statistic W^+ to be large when the alternative hypothesis is true. The approximate P-value is $P(Z \geq 2.04) = 0.021$. We conclude that the cola does lose sweetness in storage.

The output from the Minitab computer package on the next page gives a similar result. Many computer packages, including Minitab, include a correction to the standard deviation in the normal approximation to account for the ties in the ranks. This is why the P-value given by Minitab is slightly different than the one we obtained.

```
Wilcoxon Signed Rank Test

TEST OF MEDIAN = 0.000000 VERSUS MEDIAN G.T. 0.000000

                   N FOR   WILCOXON             ESTIMATED
             N     TEST    STATISTIC  P-VALUE      MEDIAN
Loss        10      10        47.5     0.023        1.150
```

b) The conclusions are the same and the *P*-values are also quite similar. The one-sample *t* test hypotheses are

H_0: $\mu = 0$

H_a: $\mu > 0$

Both tests assume that the tasters in the study are a simple random sample of all tasters. The one-sample *t* test also assumes that the before storage – after storage sweetness differences are normally distributed.

Exercise 23.39

a) The null and alternative hypotheses for the ANOVA test are

H_0: $\mu_0 = \mu_{1000} = \mu_{5000} = \mu_{10000}$

H_a: not all four means are equal

and the null and alternative hypotheses for the Kruskal-Wallis test are

H_0: seedling growths have the same distribution in all groups

H_a: seedling growth is systematically higher in some groups than in others

When the distributions have the same shape, the null hypothesis for the Kruskal-Wallis is that the median growth in all groups are equal, and the alternative hypothesis is that not all four medians are equal.

b) The medians for the nematode groups are given below. The ordered observations from the first group are 9.1, 9.2, 10.8, 13.5. The median is the average of the two middle observations, (9.2 + 10.8)/2 = 10.0.

Nematodes	Median
0	10.00
1000	11.10
5000	5.20
10000	5.55

The medians suggest nematodes reduce growth rate. There appears not to be a decrease when nematodes are at the level of 1000, but the growth drops off at 5000 and no further reduction is apparent at 10000.

It is important to note that neither the ANOVA test nor the Kruskal-Wallis test are designed specifically for the alternative that nematodes retard growth (this would be the analog of a one-sided alternative). Both procedures test the alternative of *any* differences between the groups.

The computations required for the Kruskal-Wallis test statistic are summarized in the tables below.

Growth	3.2	4.6	5.0	5.3	5.4	5.8	7.4
Group	10000	5000	5000	10000	5000	10000	5000
Rank	1	2	3	4	5	6	7

Growth	7.5	8.2	9.1	9.2	10.8	11.1	11.1
Group	10000	1000	0	0	0	1000	1000
Rank	8	9	10	11	12	13.5	13.5

```
Growth   11.3    13.5
Group    1000       0
Rank       15      16
```

Nematodes	Ranks	Sum of Ranks
0	10, 11, 12, 16	49
1000	9, 13.5, 13.5, 15	51
5000	2, 3, 5, 7	17
10000	1, 4, 6, 8	19

$$H = \frac{12}{N(N+1)} \sum \frac{R_i^2}{n_i} - 3(N+1) = \frac{12}{16(16+1)} \left(\frac{49^2}{4} + \frac{51^2}{4} + \frac{17^2}{4} + \frac{19^2}{4} \right) - 3(16+1) = 11.34$$

Since $I = 4$ groups, the sampling distribution of H is approximately chi-square with $4 - 1 = 3$ degrees of freedom. From Table E we see the P-value is approximately 0.01. There is strong evidence of a difference in seedling growth between the four groups.

The MINITAB software gives the output below when doing the Kruskal-Wallis test. The medians, average ranks (in place of sums of ranks), H statistic and P-value are given. The H statistic with an adjustment for ties in the ranks is also given.

Kruskal-Wallis Test

```
LEVEL      NOBS     MEDIAN   AVE. RANK
    1         4     10.000        12.3
    2         4     11.100        12.8
    3         4      5.200         4.2
    4         4      5.550         4.7
OVERALL      16                    8.5

H = 11.34  d.f. = 3  p = 0.010
H = 11.35  d.f. = 3  p = 0.010 (adjusted for ties)
```

CHAPTER 24

STATISTICAL PROCESS CONTROL

OVERVIEW

In practice, work is often organized into a chain of activities that lead to some result. Such a chain of activities that turns inputs into outputs is called a **process**. A process can be described by a **flow chart,** which is a picture of the stages of a process. A **cause-and-effect diagram,** which displays the logical relationships between the inputs and output of a process, is also useful for describing and understanding a process.

All processes have variation. If the pattern of variation is stable over time, the process is said to be in statistical control. In this case, the sources of variation are called **common causes.** If the pattern is disrupted by some unusual event, **special cause** variation is added to the common cause variation. **Control charts** are statistical plots intended to warn when a process is disrupted or **out of control.**

Standard **3σ control charts** plot the values of some statistic Q for regular samples from the process against the time order in which the samples were collected. The **center line** of the chart is at the mean of Q. The **control limits** lie three standard deviations of Q above (the **upper control limit**) and below (the **lower control limit**) the center line. A point outside the control limits is an **out-of-control signal.** For **process monitoring** of a process that has been in control, the mean and standard deviations used to establish the center line and control limits are based on past data and are updated regularly.

When we measure some quantitative characteristic of a process, we use $\bar{x}$ and s **charts** for process control. The $\bar{x}$ chart plots the sample means of samples of size n from the process and the s chart the sample standard deviations. The s chart monitors variation within individual samples from the process. If the s chart is in control, the $\bar{x}$ chart monitors variation from sample to sample. To interpret charts, always look first at the s chart.

For a process that is in control with mean μ and standard deviation σ, the 3σ $\bar{x}$ chart based on samples of size n has center line and control limits

$$\text{CL} = \mu,\ \text{UCL} = \mu + 3\frac{\sigma}{\sqrt{n}},\ \text{LCL} = \mu - 3\frac{\sigma}{\sqrt{n}}$$

The 3σ s chart has control limits

$$\text{UCL} = (c_4 + 2c_5)\sigma = B_6\sigma,\ \text{LCL} = (c_4 - 2c_5)\sigma = B_5\sigma$$

and the values of c_4, c_5, B_5, and B_6 can be found in Table 24.3 in your textbook for n from 2 to 10.

An ***R* chart** based on the range of observations in a sample is often used in place of an s chart. We will rely on software to produce such charts. Formulas can be found in books on quality control. $\bar{x}$ and R charts are interpreted the same way as $\bar{x}$ and s charts.

It is common to use various **out-of-control signals** in addition to "one point outside the control limits." In particular, a **runs signal** (nine consecutive points above the center line or nine consecutive points below the center line) for an $\bar{x}$ chart allows one to respond more quickly to a gradual drift in the process center.

We almost never know the mean μ and standard deviation σ of a process. These must be estimated from past data. We estimate μ by the mean $\bar{\bar{x}}$ of the observed sample means $\bar{x}$. We estimate σ by

$$\hat{\sigma} = \frac{\bar{s}}{c_4}$$

where $\bar{s}$ is the mean of the observed sample standard deviations. **Control charts based on past data** are used at the **chart setup** stage for a process that may not be in control. Start with control limits calculated from the same past data that you are plotting. Beginning with the s chart, narrow the limits as you find special causes and remove the points influenced by these causes. When the remaining points are in control, use the resulting limits to monitor the process.

Statistical process control maintains quality more economically than inspecting the final output of a process. Samples that are **rational subgroups** (subgroups that capture the features of the process in which we are interested) are important to effective control charts. A process in control is stable, so that we can predict its behavior. If individual measurements have a Normal distribution, we can give the **natural tolerances.**

A process is **capable** if it can meet or exceed the requirements placed on it. Control (stability over time) does not in itself improve capability. Remember that control describes the internal state of the process, whereas capability relates the state of the process to external specifications.

There are control charts for several different types of process measurements. One important type is the ***p* chart** which is a control chart based on plotting sample proportions $\hat{p}$ from regular samples from a process against the order in which the samples were taken. We estimate the process proportion p of "successes" by

$$\bar{p} = \frac{\text{total number of successes in past samples}}{\text{total number of opportunities in these samples}}$$

and then the control limits for a p chart for future samples of size n are

$$\text{UCL} = \bar{p} + 3\sqrt{\frac{\bar{p}(1-\bar{p})}{n}}, \text{CL} = \bar{p}, \text{LCL} = \bar{p} - 3\sqrt{\frac{\bar{p}(1-\bar{p})}{n}}$$

The interpretation of p charts is very similar to that of $\bar{x}$ charts. The out-of-control signals used are also the same as for $\bar{x}$ charts.

GUIDED SOLUTIONS

Exercise 24.1

KEY CONCEPTS - flowcharts and cause-and-effect diagrams

For this exercise, it is important to choose a process that you know well so that you can describe it carefully and recognize those factors that affect the process. Use the space provided for your flowchart and cause-and-effect diagram.

Exercise 24.4

KEY CONCEPTS - Pareto charts

What percent of total losses do these 9 DRGs account for?

Sum of percent losses =

Use the axes below to make your Pareto chart.

Which DRGs should the hospital study first when attempting to reduce its losses?

Exercise 24.7

KEY CONCEPTS - common causes

Refer to Exercise 24.1 in this Study Guide. For a process you know well, what are some common sources of variation in the process?

What are some special causes that might drive the process out of control?

Exercise 24.15

KEY CONCEPTS - $\bar{x}$ and s charts

For the first two samples in Figure 24.10 of your textbook compute $\bar{x}$ and s.

Sample 1

$\bar{x}$ =

s =

Sample 2

$\bar{x}$ =

s =

If you have access to statisical software, use the software to make your $\bar{x}$ and s charts. Otherwise, do the following.

To make the $\bar{x}$ chart, compute

$$\text{UCL} = \mu + 3\frac{\sigma}{\sqrt{n}} =$$

$$\text{CL} = \mu =$$

$$\text{LCL} = \mu - 3\frac{\sigma}{\sqrt{n}} =$$

Now plot the UCL, CL, LCL, and the values of $\bar{x}$ for all 18 sanples in the chart that follows.

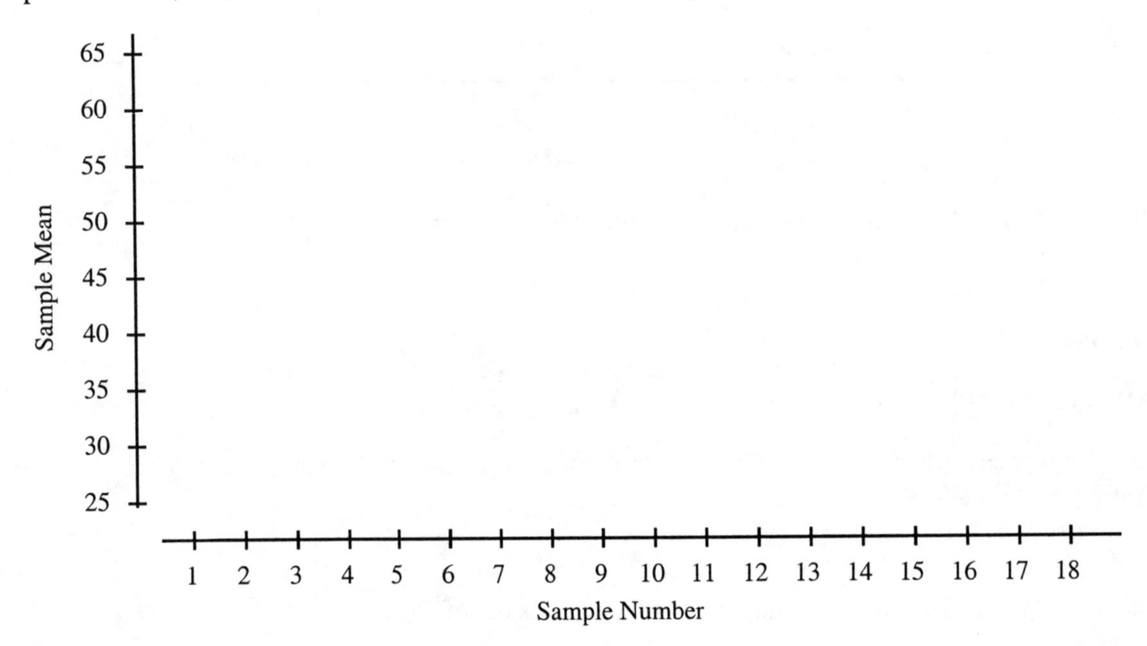

To make the s chart, compute

$$\text{UCL} = B_6\sigma =$$

$$\text{CL} = c_4\sigma =$$

$$\text{LCL} = B_5\sigma =$$

Now plot the UCL, CL, LCL, and the values of s for all 18 sanples in the chart that follows.

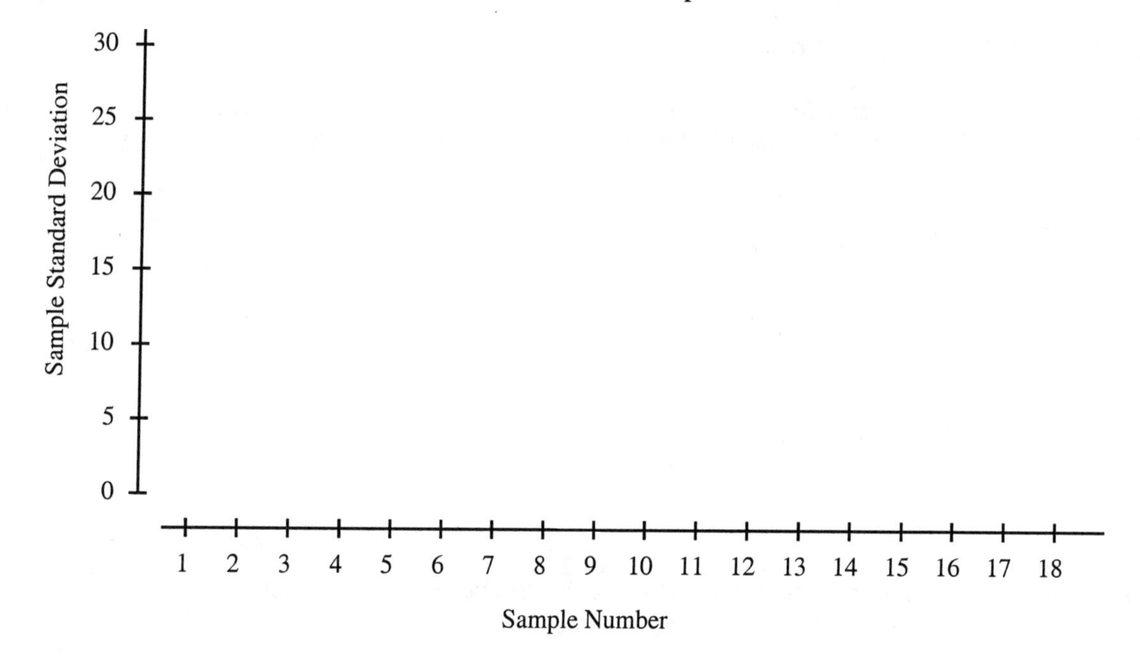

How would you describe the state of the process?

Exercise 24.20

KEY CONCEPTS - $\bar{x}$ and s control charts using past data

a) From the values of $\bar{x}$ and s in Table 24.1 of your textbook, compute (by hand, calculator, or using software)

$\bar{\bar{x}}$ = mean of the 20 values of $\bar{x}$ =

$\bar{s}$ = mean of the 20 values of s =

hence we estimate μ to be

$$\hat{\mu} = \bar{\bar{x}} =$$

and we estimate σ to be

$$\hat{\sigma} = \frac{\bar{s}}{c_4} =$$

b) Look at the s chart in Figure 24.7 of your textbook. What patterns do you see that might suggest the process σ may now be less than 43mV?

Exercise 24.29

KEY CONCEPTS - natural tolerances

The natural tolerances are $\mu \pm 3\sigma$. We do not know μ and σ, so we must estimate them from the data. We remove sample 5 from the data. Based on the remaining 17 samples, estimate

$\bar{\bar{x}}$ = mean of the 17 values of $\bar{x}$ =

$\bar{s}$ = mean of the 17 values of s =

hence we estimate μ to be

$$\hat{\mu} = \bar{\bar{x}} =$$

and we estimate σ to be

$$\hat{\sigma} = \frac{\bar{s}}{c_4} =$$

Based on these estimates, the natural tolerances for the distance between the holes are

$$\hat{\mu} \pm 3\hat{\sigma} =$$

Exercise 24.30

KEY CONCEPTS - capability

Refer to Exercise 24.29 in this Study Guide. Based on the 17 samples that were in control, we saw in Exercise 24.29 of this Study Guide that estimates of μ and σ are $\hat{\mu} = 43.41$ and $\hat{\sigma} = 12.39$. We therefore assume that distances between holes vary from meter to meter according to an $N(43.41, 12.39)$ distribution. Use normal probability calculations to find the probability that the distance x between holes in a randomly selected meter is between 54 ± 10 (i.e., between 44 and 64). Refer to Chapter 3 of your textbook if you have forgotten how to do normal probability calculations.

$$P(44 < x < 64) =$$

We conclude that about what percent of meters meet specifications?

Exercise 24.34

KEY CONCEPTS - p charts

To find the appropriate center line and control limits, we must first compute $\bar{p}$. The total number of opportunities for missing or deformed rivets is just the total number of rivets, because each rivet has the possibility of being missing or deformed. The number of "successes" in past samples is just the missing or deformed rivets in the recent data. What are these values? Now estimate $\bar{p}$.

$$\bar{p} = \frac{\text{total number of successes in past samples}}{\text{total number of opportunities in these samples}} =$$

The next wing contains n = 1070 rivets, and the control limits for a p chart for future samples of size n = 1070 are

$$\text{UCL} = \bar{p} + 3\sqrt{\frac{\bar{p}(1-\bar{p})}{n}} =$$

$$\text{CL} = \bar{p} =$$

$$\text{LCL} = \bar{p} - 3\sqrt{\frac{\bar{p}(1-\bar{p})}{n}} =$$

Complete Solutions

Exercise 24.1

We take as our example the process of making a cup of coffee. A possible flow chart and cause-and-effect diagram of the process follow.

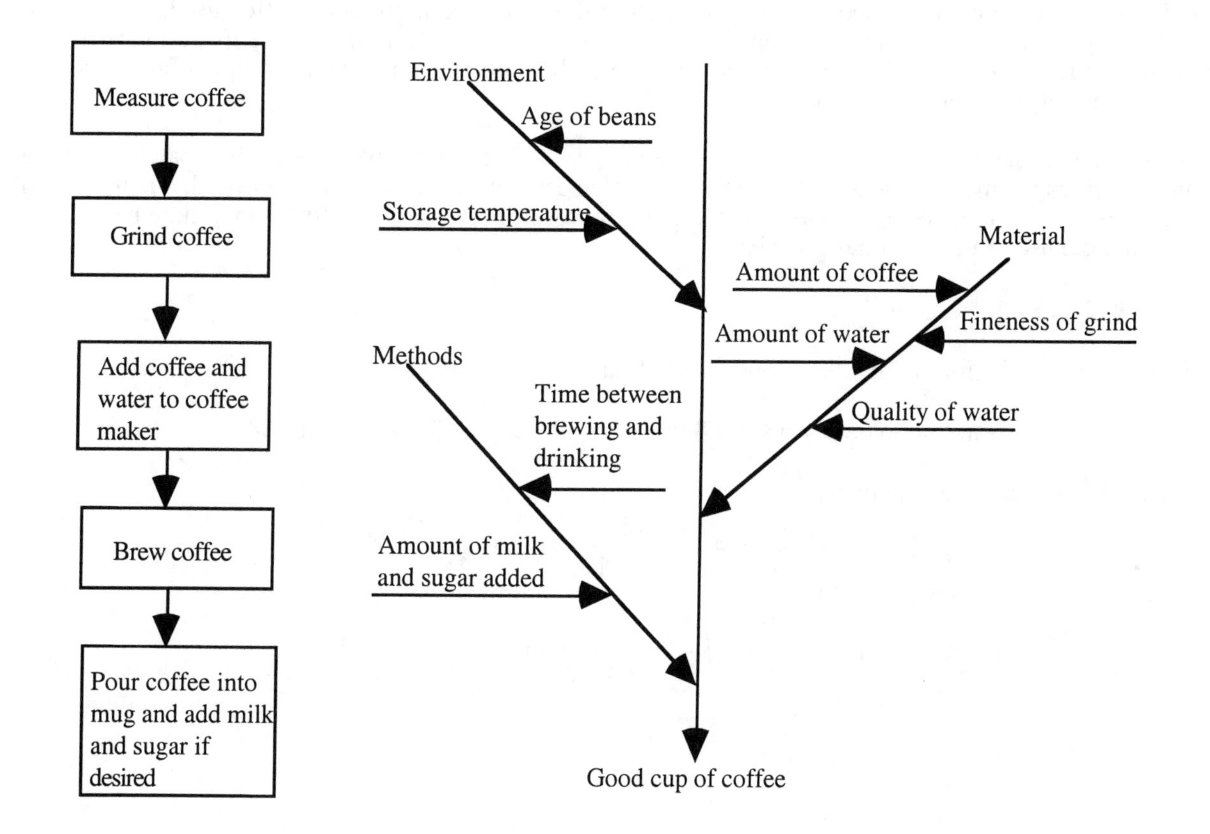

Exercise 24.4

Adding the percents listed, one finds that the percent of total losses that these 9 DRGs account for is 80.5%. A Pareto chart of losses by DRG follows.

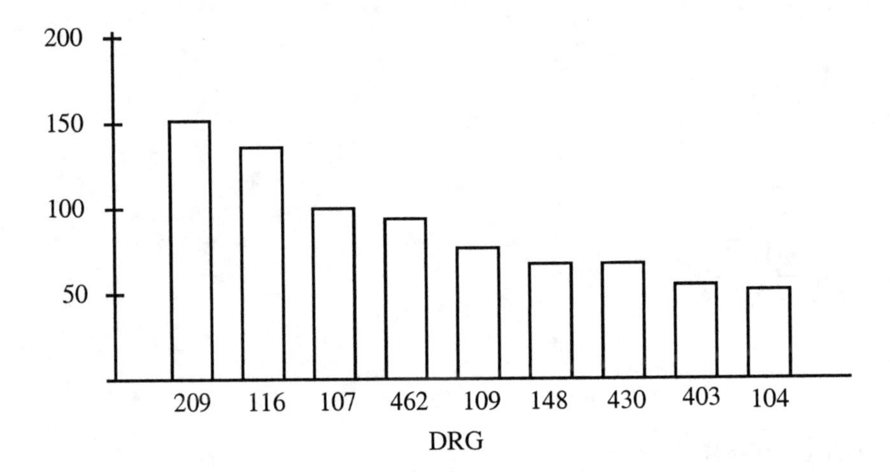

The hospital ought to study DRGs 209 and 116 first in attempting to reduce its losses. These are the two DRGs with the largest percent losses and combined account for nearly 30% of all losses.

Exercise 24.7

In Exercise 24.1 of this Study Guide, we described the process of making a good cup of coffee. Some sources of common-cause variation are variation in how long the coffee has been stored and the conditions under which it has been stored, variation in the measured amount of coffee used, variation in how finely ground the coffee is, variation in the amount of water added to the coffee maker, variation in the length of time the coffee sits between when it has finished brewing and when it is drunk, and variation in the amount of milk and/or sugar added.

Some special causes that might at times drive the process out of control would be a bad batch of coffee beans, a serious mismeasurement of the amount of coffee used or the amount of water used, a malfunction of the coffee maker or a power outage, interruptions that result in the coffee sitting a long time before it is drunk, and the use of milk that has gone bad.

Exercise 24.15

We compute $\bar{x}$ and s for the first two samples and find

$$\text{First sample: } \bar{x} = 48,\ s = 8.94;\ \ \text{Second sample: } \bar{x} = 46,\ s = 13.03$$

To make the $\bar{x}$ chart, we note that

$$\text{UCL} = \mu + 3\frac{\sigma}{\sqrt{n}} = 43 + 3\frac{12.74}{\sqrt{5}} = 43 + 17.09 = 60.09$$

$$\text{CL} = \mu = 43$$

$$\text{LCL} = \mu - 3\frac{\sigma}{\sqrt{n}} = 43 - 3\frac{12.74}{\sqrt{5}} = 43 - 17.09 = 25.91$$

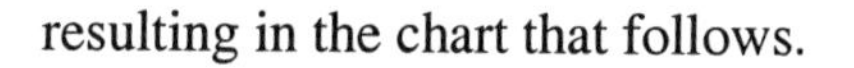

resulting in the chart that follows.

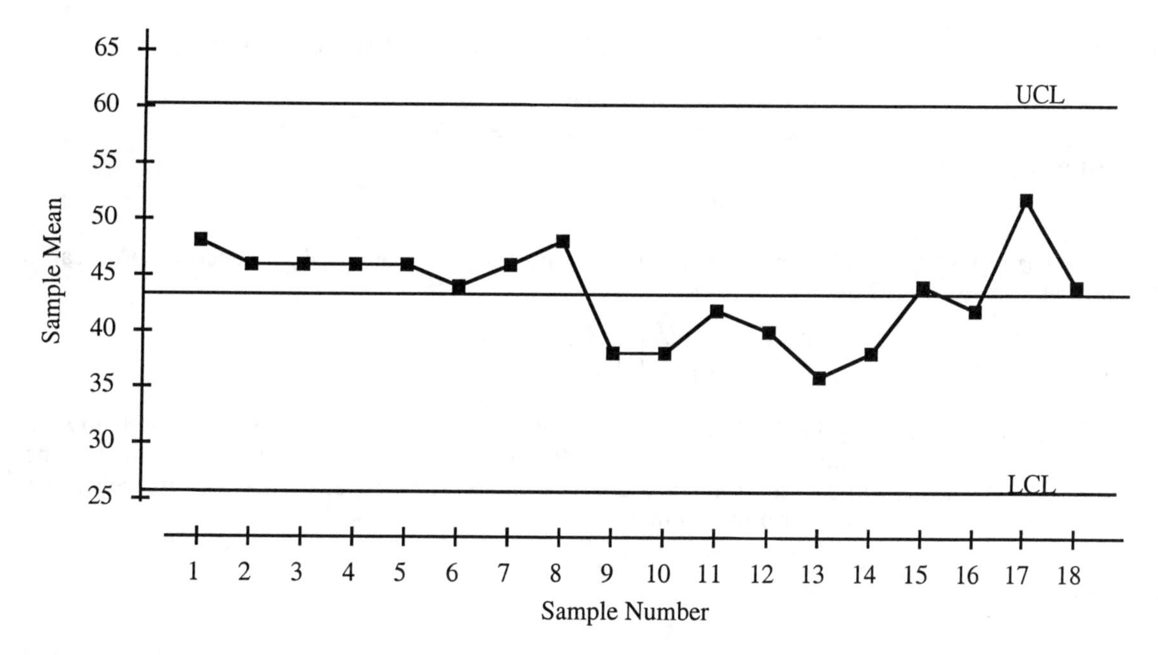

To make the s chart, we note that

$$\text{UCL} = B_6\sigma = 1.964(12.74) = 25.02$$

$$\text{CL} = c_4\sigma = 0.9400(12.74)) = 11.98$$

$$\text{LCL} = B_5\sigma = 0(12.74)) = 0$$

resulting in the chart that follows.

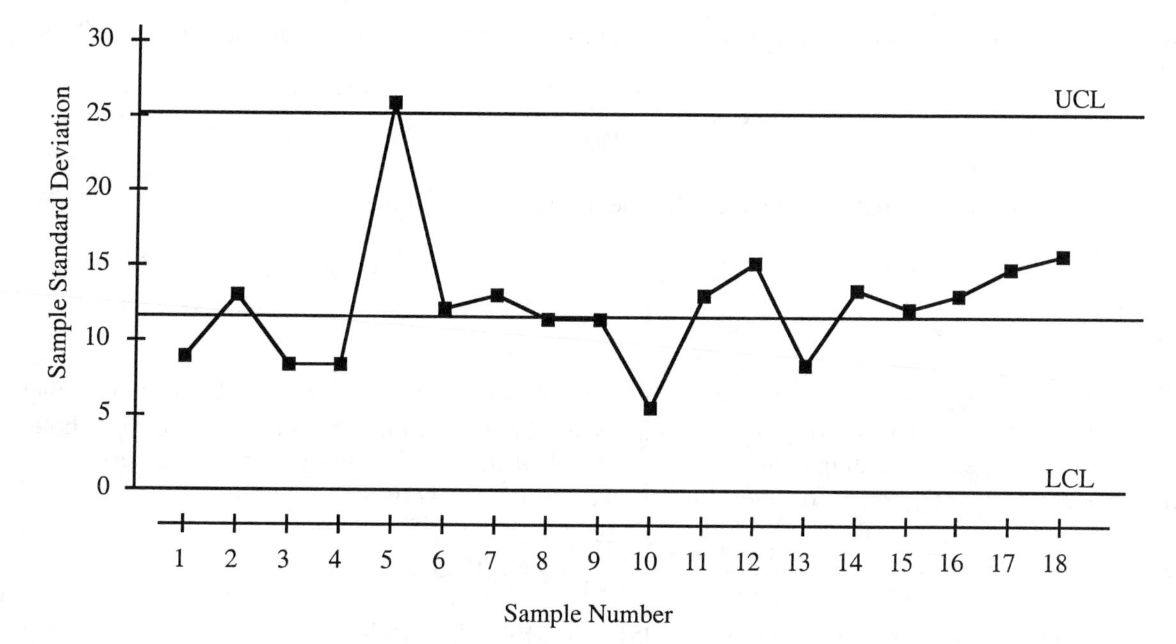

The s chart shows a lack of control at sample point 5, but otherwise neither chart shows a lack of control. We would want to find out what happened at sample 5 to cause a lack of control in the s chart.

Exercise 24.20

a) From the values of $\bar{x}$ and s in Table 24.1 of your textbook, we compute (using software)

$$\bar{\bar{x}} = \text{mean of the 20 values of } \bar{x} = 275.065$$
$$\bar{s} = \text{mean of the 20 values of } s = 34.55$$

hence we estimate μ to be

$$\hat{\mu} = \bar{\bar{x}} = 275.065$$

and we estimate σ to be (using the fact that the samples are each of size $n = 4$ and according to Table 24.3 of your textbook, $c_4 = 0.9213$)

$$\hat{\sigma} = \frac{\bar{s}}{c_4} = \frac{34.55}{0.9213} = 37.5$$

b) If we look at the s chart in Figure 24.7 of your textbook we see that most of the points lie below 40 (and more than half of those below 40 lie well below 40), while of the points above 40, all but one (sample 12) are only slightly larger than 40. The s chart suggests that typical values of s are below 40, which is consistent with the estimate of σ in part (a).

Exercise 24.29

The natural tolerances are $\mu \pm 3\sigma$. We do not know μ and σ, so we must estimate them from the data. We remove sample 5 from the data. Based on the remaining 17 samples, we find

$$\bar{\bar{x}} = \text{mean of the 17 values of } \bar{x} = 43.41$$
$$\bar{s} = \text{mean of the 17 values of } s = 11.65$$

hence we estimate μ to be

$$\hat{\mu} = \bar{\bar{x}} = 43.41$$

and we estimate σ to be (using the fact that the samples are each of size $n = 5$ and according to Table 24.3 of your textbook, $c_4 = 0.9400$)

$$\hat{\sigma} = \frac{\bar{s}}{c_4} = \frac{11.65}{0.9400} = 12.39$$

Based on these estimates, the natural tolerances for the distance between the holes are

$$\hat{\mu} \pm 3\hat{\sigma} = 43.41 \pm 3(12.39) = 43.41 \pm 37.17 \text{ or } 6.24 \text{ to } 80.58.$$

Exercise 24.30

Based on the 17 samples that were in control, we saw in Exercise 24.29 in this Study Guide that estimates of μ and σ are $\hat{\mu} = 43.41$ and $\hat{\sigma} = 12.39$. We therefore assume that distances between holes vary from meter to meter according to an $N(43.41,12.39)$ distribution. The probability that the distance x between holes in a randomly selected meter is between 54 ± 10 (i.e., between 44 and 64) is thus

$$P(44 < x < 64) = P\left(\frac{44 - 43.41}{12.39} < \frac{x - 43.41}{12.39} < \frac{64 - 43.41}{12.39}\right) = P(0.05 < Z < 1.66)$$
$$= P(Z < 1.66) - P(Z < 0.05) = 0.9515 - 0.5199 = 0.4316$$

We conclude that about 43.16% of meters meet specifications.

Exercise 24.34

The total number of opportunities for missing or deformed rivets is just the total number of 34700 rivets, because each rivet has the possibility of being missing or deformed. The number of "successes" in past samples is just the 208 missing or deformed rivets in the recent data. We therefore estimate the process proportion p of "successes" from the recent data by

$$\bar{p} = \frac{\text{total number of successes in past samples}}{\text{total number of opportunities in these samples}} = \frac{208}{34700} = 0.00599$$

The next wing contains n = 1070 rivets, and the control limits for a p chart for future samples of size n = 1070 are

$$\text{UCL} = \bar{p} + 3\sqrt{\frac{\bar{p}(1-\bar{p})}{n}} = 0.00599 + 3\sqrt{\frac{0.00599(1-0.00599)}{1070}} = 0.00599 + 0.00708 = 0.01307$$

$$\text{CL} = \bar{p} = 0.00599$$

$$\text{LCL} = \bar{p} - 3\sqrt{\frac{\bar{p}(1-\bar{p})}{n}} = 0.00599 - 3\sqrt{\frac{0.00599(1-0.00599)}{1070}} = 0.00599 - 0.00708 = 0$$

Note that in the LCL, we set negative values to 0 because a proportion can never be less than 0.